Gender and Development

Janet Henshall Momsen

Routledge
Taylor & Francis Group

LONDON AND NEW YORK

First published 2004
by Routledge
11 New Fetter Lane, London EC4P 4EE

Simultaneously published in the USA and Canada
by Routledge
29 West 35th Street, New York, NY 10001

Routledge is an imprint of the Taylor & Francis Group

© 2004 Janet Henshall Momsen

Typeset in Times by
Florence Production Ltd, Stoodleigh, Devon
Printed and bound in Great Britain by
The Cromwell Press, Trowbridge, Wiltshire

British Library Cataloguing in Publication Data
A catalogue record for this book is available from the British Library

Library of Congress Cataloging in Publication Data
Momsen, Janet Henshall.
Gender and development / Janet Henshall Momsen.
p. cm. – (Routledge perspectives on development)
Includes bibliographical references and index.
1. Women in development – Developing countries.
2. Sexual division of labor – Developing countries.
3. Women – Developing countries. I. Title. II. Series.
HQ1240.5.D44M657 2004
305.3′09172′4 – dc21   2003008539

ISBN 0–415–26689–0 (hbk)
ISBN 0–415–26690–4 (pbk)

Contents

Plates

Figures

Tables

Boxes

Acknowledgements

This book owes much to the encouragement and shared knowledge offered by my friends and colleagues, members of the International Geographical Union's Commission on Gender, especially during our travels and conferences in many different countries. I also learned valuable lessons about development policy from those who served with me on the Board of the Association of Women and Human Rights in Development (AWID).

I am grateful to my students for sharing their insights with me, both in the classroom and from their research. I am also indebted to my students and colleagues who gave me permission to use their photographs and quote from their research in this volume. I especially thank Margareta Lelea who made the maps for me and tracked down innumerable references in the library. I also thank my editor, Andrew Mould, and his assistants, for their patience and help.

Above all I am grateful to all those people in developing countries who allowed me to take up their time with questions and who taught me so much. As usual, I am to blame for the mistakes which may remain.

The author and publishers would like to thank the following for granting permission to reproduce material in this work: Seela Aladuwaka for Plates 8.8, 8.9 and 8.10; Michael Appel for Plate 5.5; Mariamba Awumbila for Box 9.1 and Table 9.1; Jane Benton for Box 7.1; Amriah Buang for Box 8.2; Vincent Dao for Plates 1.2,

6.7 and 8.4; Allison Griffith for Table 5.1; Colette Harris for Box 4.1; Indra Harry for Table 6.1; Shahnaz Huq-Hussain for Box 7.2; The International Women's Tribute Centre for Figure 3.4; Margareta Lelea for Plate 6.13; Janice Monk Plate 6.11; Claudel Noel for Tables 7.1, 7.2 and 7.3; Emily Oakley for Plate 6.8; Jeanine Pfeiffer for Figure 6.2; Vidyamali Samarasinghe for Box 6.1; Garrett Smith for Table 6.3; Rebecca Torres for Plates 6.1 and 8.6; and Janet Townsend for Plates 5.1, 6.3 and 7.1.

1 Introduction: gender is a development issue

Learning objectives

When you have finished reading this chapter, you should be able to:

- understand flexible gender identities and roles
- appreciate the gender impact of sudden economic change
- be aware of different approaches to gender and development
- be familiar with the basic spatial patterns of gender and development.

The development process affects women and men in different ways. (The after effects of colonialism, and the peripheral position of poor countries of the South and those with economies in transition in today's globalizing world, exacerbate the effects of discrimination on women.)The penetration of capitalism, leading to the modernization and restructuring of subsistence and centrally planned economies, often increases the gender-based disadvantages. The modern sector takes over many of the economic activities, such as food processing and making of clothes, which had long been the means by which women supported themselves and their families. But by relieving them of these time consuming chores it gives them the freedom to find other, perhaps better, sources of earned income. Yet a majority of the better-paid jobs involving new technology go to men, but male income is less likely to be spent on the family.

Modernization of agriculture has altered the division of labour between the sexes, increasing women's dependent status as well as their workload. Women often lose control over resources such as land and are generally excluded from access to improved agricultural methods. Male mobility is higher than female, both between places and between jobs, and more women are being left alone to support children. In some countries, especially in the Middle East, South Asia and Latin America, women cannot do paid work or travel without their husband's or father's written permission. Women carry a double or even triple burden of work as they cope with housework, childcare and subsistence food production, in addition to an expanding involvement in paid employment. Everywhere women work longer hours than men. The pressure on gender relations of the changing status of women, and of rapid economic restructuring combined with growing impoverishment at the household level for many, is crucial to the success or failure of development policies.

Gender (the socially acquired notions of masculinity and femininity by which women and men are identified) is a widely used and often misunderstood term. It is sometimes conflated with sex or used to refer only to women. Gender identities, because they are socially acquired and based on nurture, vary. In Polynesia gender identities are often flexible. In families without daughters, one son is selected when very young to be raised as a girl to fulfil the family's needs for someone to undertake a daughter's roles, such as care of siblings and housework. As adults, these individuals usually continue to live and dress as women, and occupy female roles with jobs as waitresses or maids in the rapidly growing tourist industry, or even as transvestite prostitutes. Today the *faafafine* (trying to be like a lady) are also found in Melanesia and are becoming more open and in some forms more aggressive (Fairbairn-Dunlop 2002). In Western Samoa they often work as dressmakers and school teachers and may run drag queen contests and fund raisers for church groups (ibid. 2002).

Gender relations (the socially constructed form of relations between women and men) have been interrogated in terms of the way development policies change the balance of power between women and men. Gender roles (the household tasks and types of employment socially assigned to women and men) are not fixed and globally consistent and indeed become more flexible with the changes brought about by economic development. Everywhere gender is crosscut by differences in class, race, ethnicity, religion and age. The much-criticized binary division between 'Western' women and the 'Other', between white and non-white and between colonizer and colonized is both patronizing and simplistic (Mohanty 1984). Feminists have often seen women as socially constituted as a homogeneous group on the basis of shared oppression. But in order to understand these gender relations we must interpret them within specific societies and on the basis of historical and political practice, not a priori on gender. Different places and societies have different practices and it is necessary to be cognizant of this heterogeneity within a certain global homogeneity of gender roles. At the same time we need to be aware of different voices and to give them agency. The subaltern voice is hard to hear but by presenting experiences from fieldwork I have tried to incorporate it. The voices of educated women and men of the South can interpret postcoloniality but because they write in the colonizers' languages their voices have to be listened to on several levels. By combining an appreciation of different places and different voices we can arrive at an understanding of how the process of economic change in the South and in the post-communist countries is impacting people and communities (Kinnaird and Momsen 1993).

Clearly, the roles of men and women in different places show great variation: most clerks in Martinique are women but this is not so in Madras, just as women make up the vast majority of domestic servants in Lima but not in Lagos. Nearly 90 per cent of sales workers in Accra are women but in Algeria they are almost all men (Plates 1.1 and 1.2). In every country the jobs done predominantly by women are the least well paid and have the lowest status. In the countries of Eastern and Central Europe, Russia and China, where most jobs were open to men and women under communism, the transition to capitalism has led to increased unemployment, especially for women, except in Hungary, where the particular character of gendered education and employment resulted in more men's jobs being lost. In most parts of the world the gender gap in political representation has become smaller but in the former USSR and its satellite countries in Eastern and Central Europe there has been a rapid decline in average female representation in parliament from 27 per cent in 1987 under communism, to 7 per cent in 1994 (United Nations 1995b). This has been most marked in Romania, where the figures were 34 per cent in 1987 and 4 per cent in 1994 (United Nations 1995b) rising to 5.6 per cent in 2000 (Elson 2000). The relationship between development and the spatial patterns of the gender gap provides the main theme of this book.

Plate 1.1 *Egypt: Cairo market with men selling fresh vegetables and lemons and women with children shopping. The women are covered from head to toe in this public space.*

Source: author

At the beginning of the third millennium most of the world's population is living more comfortably than it was a century ago. Women as a group now have a greater voice in both their public and private lives. The spread of education and literacy has opened up new opportunities for many people and the time–space compression associated with globalization is making possible the increasingly rapid and widespread distribution of information and scientific knowledge. Improvements in communications, however, also make us aware that economic development is not always unidirectional and benefits are not equally available. For the Afghan women living under the Taliban regime, forced to be enshrouded in burkas and refused the right to work or go to school, it was hard. 'We were like

Plate 1.2 *Burkina Faso: women vegetable growers accompanied by small children, selling their produce in the market in the town of Ouahigouya. Buyers come from as far away as Togo to this market.*

Source: Vincent Dao, University of California, Davis

in prison. We had no life, nothing for us to do. We were not people', according to a hairdresser reopening her beauty salon after the fall of Kabul in November 2001 (Gannon 2001: 27). Unfortunately, outside Kabul not much has changed – Afghan women still wear burkas and few girls are able to attend school, although 20 years ago Afghanistan had a very cultured society with many highly educated women and men. In those days few women were veiled and most had considerable freedom of movement.

Women's organizations, and the various United Nations international women's conferences in Mexico City, Copenhagen, Nairobi and Beijing over the last three decades, have put gender issues firmly on the development agenda but economic growth and modernization is not gender neutral. The experiences of different states and regions show that economic prosperity helps gender equality but some gender gaps are resistant to change. Rapid growth, as in the East Asian countries, has led to a narrowing of the gender differences in wages and education but inequality in political representation

Box 1.1

Economic crisis leads to setbacks for women in Argentina

One out of five couples in Argentina experiences domestic violence, 400,000 illegal abortions are performed annually, accounting for 29 per cent of maternal deaths, and women earn 40 per cent less than men with similar educational levels, according to a new report presented to CEDAW. The report, drawn up by seven women's and human rights groups, says that the country's severe economic crisis at the start of the new millennium has aggravated problems like domestic violence, teen pregnancy and health care.

Signatories to CEDAW are required to implement policies aimed at enforcing the rights of women and must provide information to CEDAW every four years so that progress, setbacks and compliance with policies can be assessed. The report by the seven civil society organizations is intended to complement the official information provided by the Argentinian government. It points out that virtually none of the CEDAW committee's recommendations to the Argentine state in 1997 have been met.

An impact of the Argentinian economic crisis has been the decline of influence of the National Women's Council, which was 'demoted' to a mere programme with consequent loss of funding in January 2002. Despite CEDAW's criticism in 1997 that the state was not doing enough to address the growing phenomenon of prostitution and sexual exploitation of girls, nothing has been done to tackle the problem. In fact, the 2002 civil society report notes that the economic crisis has triggered a significant rise in prostitution among women and girls due to the tremendous impoverishment of families, and networks trafficking in women and girls have expanded. Access to health services has been severely curtailed and availability of contraceptives has declined. Consequently, women are being forced increasingly to turn for family planning to abortion, which is illegal in Argentina and so often dangerous. The economic crisis has led to the sexual exploitation of younger and younger girls so that, for the first time, pregnancy-related deaths of girls under the age of 15 are occurring. Overall, the report to CEDAW notes that women and children are the main victims of the growing poverty in Argentina, which now affects 51.4 per cent of the population.

Source: adapted from InterPress Service, 6 August 2002.
Available e-mail: awid@awid.org (17 August 2002).

remains. Sudden, economic change, such as structural adjustment programmes or the post-cold war transition in Eastern Europe, creates new gender differences in which women are generally the losers (Box 1.1).

Authorial positionality

As a Western white woman feminist writing about women and men in the developing and transitional countries there is clearly a huge gap between observer and observed. As a dual national (British and Canadian) currently teaching in the USA, who has lived and taught in the Caribbean, Costa Rica, Brazil and Nigeria, and carried out fieldwork in such disparate areas as the mountains of southern China among minority Yi people, with Mayans in Mexico and in Hungarian villages, over 40 years' experience has taught me a lot. I have also benefited enormously from working with wise colleagues and graduate students from developing countries including Bangladesh, Barbados, Brazil, Burkina Faso, Ghana, Hungary, India, Jamaica, Lesotho, Libya, Nigeria, Singapore, Sri Lanka, St Kitts-Nevis, Trinidad and Western Samoa, and with fellow members of the Board of the Association of Women and Human Rights in Development (AWID). Above all, the award of the position of an honorary Queen Mother, with the title of Nana Ama Sekiybea, by the Chief of an Akuapem village in southern Ghana was especially meaningful.

Development

After the Second World War, the United States and its allies recognized the need for a programme that would spread the benefits of scientific and industrial progress. In this way two-thirds of the world was defined as underdeveloped, foreign aid became an accepted but declining part of national budgets and development agencies began to proliferate. Gradually foreign aid, including food aid and military aid, became a political tool used by the superpowers of the USA and the USSR in a cold war competition to influence the ex-colonial and non-aligned nations of the so-called 'Third World'. With the collapse of the state socialist model in the USSR and Eastern Europe in 1989, the American model of neoliberal capitalism became dominant. Although some countries, such as Cuba, China and North Korea, continued with centrally planned state socialism in some form, they either instituted market-oriented reforms, or were so damaged by the loss of financial support from Russia and the overwhelming political and economic power of the USA, that they no longer offered a viable alternative development model. Some people saw this as marking the end of development and that development had failed. Others saw it as an opportunity for rethinking and broadening the idea of development.

For many people the last few decades have brought about better living conditions, health and well-being (United Nations 1995b and 2000). Today the focus is less on increasing gross domestic product and spreading modernization, and more on emphasizing debt relief, reducing corruption, recognizing the importance of social as well as human capital and the overall reduction of poverty and disease. These development goals will be considered in terms of gender differences.

Gender equality does not necessarily mean equal numbers of men and women or girls and boys in all activities, nor does it mean treating them in the same way. It means equality of opportunity and a society in which women and men are able to lead equally fulfilling lives. The aim of gender equality recognizes that men and women often have different needs and priorities, face different constraints and have different aspirations. Above all, the absence of gender equality means a huge loss of human potential and has costs for both men and women and also for development.

Over half a century ago, in 1946, the United Nations set up the Commission on the Status of Women. It was to have two basic functions: to 'prepare recommendations and reports to the Economic and Social Council on promoting women's rights in political, economic, civil, social and educational fields'; and to make recommendations on 'urgent problems requiring immediate attention in the field of women's rights' (United Nations 1996: 13). The remit of the Commission remained essentially the same until 1987 when it was expanded to include advocacy for equality, development and peace and monitoring of the implementation of measures for the advancement of women at regional, sectoral, national and global levels (United Nations 1996). Today it is clear that progress towards gender equality in most parts of the world is considerably less than that which was hoped for. However, disparities between women in different countries are greater than those between men and women in any one country. At the beginning of the new millennium life expectancy at birth for women varies from 82 years in Hong Kong to 38 in Zambia, while male life expectancy is lower, ranging from 37 years in Angola and Zambia to 77 in Hong Kong, the same as in Sweden (PRB 2002). Globally, only 69 per cent of women but 83 per cent of men over 15 years of age are literate (PRB 2002). The proportion of illiterates in the female population varies from 92 per cent in Niger to less than 1 per cent in Barbados and Tajikistan, but in some countries, such as Lesotho, Jamaica, Uruguay, Qatar and the United Arab Emirates, a higher proportion of women than men are literate (PRB 2002). Even within individual countries women are not a homogeneous group but can be differentiated by class, race, ethnicity, religion and life stage. The elite and the young are more likely to be educated everywhere, increasing the generational gap. The range on most socio-economic measures is wider for women than for men and is greatest among the countries of the South.

As we enter the new millennium the development focus is on alleviating world poverty. The empowerment of women and the promotion of gender equality is one of the eight internationally agreed Millennium Development Goals (MDGs) designed to achieve this (Box 1.2). There is a great deal of evidence drawn from comparisons at the national and sub-national scale that societies that discriminate on the basis of gender pay a price in more poverty, slower growth and a lower quality of life, while gender equality enhances development. For example it has been estimated that increasing the education and access to inputs of female farmers relative to male farmers in Kenya would raise yields by as much as one-fifth. Literate mothers have better-fed children who are more likely to attend school. Yet in no country in the developing world do women enjoy equality with men in terms of political, legal, social and economic rights. In general, women in Eastern Europe have the greatest equality of rights, but this has declined in the last decade. The lowest equality of rights is found in South Asia, sub-Saharan Africa, the Middle East and North Africa. There are no global comparative data on rights more recent than 1990 but there is some evidence that equality of rights has improved since the 1995 Fourth World Conference on Women held in Beijing. The Convention on the Elimination of All forms of Discrimination against Women (CEDAW) was established in 1979 and came into force in 1981 after it had been ratified by 20 countries (Elson 2000). By 1996, 152 countries had become party to the Convention but in 2002 the United States had still not ratified it. Unfortunately, ratification of CEDAW

Box 1.2

Millennium Development Goal 3

Goal: to promote gender equality and empower women

Target: to eliminate gender disparities in primary and secondary education, preferably by 2005 and in all levels of education no later than 2015.

Indicators:

- Ratio of girls to boys in primary, secondary and tertiary education.
- Ratio of literate females to males of 15–24 years of age.
- Share of women in paid employment in the non-agricultural sector.
- Proportion of seats held by women in national parliaments.

Studies in many countries have shown that education for girls is the single most effective way of reducing poverty, although by itself not sufficient. In this context, the elimination of gender disparities in education has been selected as the key target to demonstrate progress towards Millennium Goal 3. However, progress towards gender equality in education is dependent on success in tackling inequalities in wider aspects of economic, political, social and cultural life and this is reflected in the indicators listed above. As in Development Goal 3, each Millennium Goal involves several indicators on which success or failure can be measured.

Adapted from Derbyshire (2002: 7).

---

does not necessarily lead to an immediate reduction in gender discrimination but it does enforce regular reporting on progress (Box 1.1).

By the turn of the century there had been three United Nations Development Decades, while the Decade for Women (1976–85) culminated in a conference in Nairobi in 1985. At the conclusion of the first two Development Decades it was found that the extent of poverty, disease, illiteracy and unemployment in the South had increased. During the 1980s we witnessed unprecedented growth of developing country debt and acute famine in Africa. Similarly the Decade for Women saw only very limited changes in patriarchal attitudes, that is institutionalized male dominance, and few areas where modernization was associated with a reversal of the overwhelming subordination of women.

Yet despite the apparent lack of change, the United Nations Decade for Women achieved a new awareness of the need to consider women when planning for development. In the United States the Percy Amendment of 1973 ensured that women had to be specifically included in all projects of the Agency for International Development. The British Commonwealth established a Woman and Development programme in 1980 supported by all member countries. In many parts of the South, women's organizations and networks at the community and national level have come to play an increasingly important role in the initiation and implementation of development projects. Above all, the Decade for Women brought about a realization that data collection and research were needed in order to document the situation of women throughout the world. The consequent outpouring of information has made this book possible.

Women and development

Prior to 1970, when Esther Boserup published her landmark book on women and development, it was thought that the development process affected men and women in the same way. Productivity was equated with the cash economy and so most of women's work was ignored. When it became apparent that economic development did not automatically eradicate poverty through trickle-down effects, the problems of distribution and equality of benefits to the various segments of the population became of major importance in development theory. Research on women in developing countries challenged the most fundamental assumptions of international development, added a gender dimension to the study of the development process and demanded a new theoretical approach.

The early 1970s' approach of 'integration', based on the belief that women could be brought into existing modes of benevolent development without a major restructuring of the process of development, has been the object of much feminist critique. The alternative vision put forward, of development *with* women, demanded not just a bigger piece of someone else's pie, but a whole new dish, prepared, baked and distributed equally. It soon became clear that a focus on women alone was inadequate and that a gendered view was needed. Women and men are affected differently by economic change and development and thus an active public policy is needed to intervene in order to close gender gaps. In the mission statement of the Beijing Fourth World Conference on Women, held in 1995, it was said that '[a] transformed partnership based on equality between women and men is a condition for people-centred sustainable development' (United Nations 1996: 652).

The focus on gender in development policies came first from the major national and international aid agencies, and governments in the South quickly learned that they needed to build a gender aspect into their requests for assistance. Thus at the beginning it was the North that largely imposed the agenda. As non-governmental organizations (NGOs) began to play an increasingly important role in grassroots delivery of aid, their gender policies began to influence local action. In a Ghanaian village the men had had a reafforestation project for a decade at some distance from the village. The chief told me that he decided to set up a women's agroforestry project, under the leadership of his sister, because he knew, from radio reports of the current interests of NGOs, that it would be easier to get outside assistance for such a project than for one involving men. In this case the agenda set at the top was manipulated from the grassroots.

Approaches to women, gender and development

By the end of the twentieth century all approaches to development involving a focus on women had been amalgamated into a gender and development (GAD) approach. Kate Young argues that this bears little similarity to the original formulation of GAD and that the term gender is often used as a mere synonym for women/woman (Young 2002). The study of masculinities and of men as the missing half of GAD is now on the agenda, but is provoking much ambivalence since it has a number of important implications for GAD policies and practice, especially in terms of undermining efforts to help women, as gender equality is still far from being achieved (Cornwall and White 2000).

Chronology of approaches

1 *The welfare approach*   Until the early 1970s development policies were directed at women only in the context of their roles as wives and mothers, with a focus on mother and child health and on reducing fertility. It was assumed that the benefits of macroeconomic strategies for growth would automatically trickle down to the poor,

and that poor women would benefit as the economic position of their husbands improved.

Boserup (1970) challenged these assumptions, showing that women did not always benefit as the household head's income increased and that women were increasingly being associated with the backward and traditional and were losing status.

2 *The WID approach* The rise of the women's movement in Western Europe and North America, the 1975 UN International Year for Women and the International Women's Decade (1976–85) led to the establishment of women's ministries in many countries and the institutionalization of Women in Development (WID) policies in governments, donor agencies and NGOs. The aim of WID was to integrate women into economic development by focusing on income generation projects for women.

This anti-poverty approach failed on its own terms as most of its income-generation projects were only marginally successful, often because they were set up on the basis of a belief that women of the South had spare time available to undertake these projects. It left women out of the mainstream of development and treated women identically. It also ghettoized the WID group within development agencies.

By the 1980s WID advocates shifted from exposing the negative effects of development on women to showing that development efforts were losing out by ignoring women's actual or potential contribution.

3 *Gender and Development (GAD)* This approach originated in academic criticism starting in the mid 1970s in the UK (Young 2002: 322). Based on the concept of gender (the socially acquired ideas of masculinity and femininity) and gender relations (the socially constructed pattern of relations between men and women) they analysed how development reshapes these power relations. Drawing on feminist political activism, gender analysts explicitly see women as agents of change. They also criticize the WID approach for treating women as a homogeneous category and they emphasize the important influence of differences of class, age, marital status, religion and ethnicity or race on development outcomes.

Proponents distinguished between 'practical' gender needs, that is items that would improve women's lives within their existing roles, and 'strategic' gender needs that seek to increase women's ability to take on new roles and to empower them (Molyneux 1985; Moser 1993). Gender analysts demanded a commitment to change in the structures of power in national and international agencies (Derbyshire 2002).

4 *Women and Development (WAD)*   At the 1975 UN Women's World Conference in Mexico City the feminist approaches of predominantly white women from the North aimed at gender equality were rejected by many women in the South who argued that the development model itself lacked the perspective of developing countries. They saw overcoming poverty and the effects of colonialism as more important than equality. Out of this grew the DAWN Network, based in the South, which aimed to make the view of developing countries more widely known and influential (Sen and Grown 1987).

By 1990 WID, GAD and WAD views had largely converged (Rathgeber 1990) but different approaches to gender and development continued to evolve.

5 *The efficiency approach*   The strategy under this approach was to argue that, in the context of structural adjustment programmes (SAPs), gender analysis made good economic sense. It was recognized that understanding men's and women's roles and responsibilities as part of the planning of development interventions improved project effectiveness. The efficiency approach was criticized for focusing on what women could do for development rather than on what development could do for women.

6 *The empowerment approach*   In the 1980s, empowerment was regarded as a weapon for the weak, best wielded through grassroots and participatory activities (Parpart 2002). However, empowerment has many meanings and by the mid 1990s some mainstream development agencies had begun to adopt the term. For the most part these institutions see empowerment as a means for enhancing efficiency and productivity without changing the status quo. The alternative development literature, on the other hand, looks to empowerment as a method of social transformation and achieving gender equality. Jo Rowlands (1997) sees empowerment as a broad development process that enables people to gain self-confidence and self-esteem, so allowing both men and women to actively participate

in development decision-making. The empowerment approach was also linked to the rise of participatory approaches to development and often meant working with women at the community level building organizational skills.

7 *Gender and the Environment (GED)*   This approach was based on ecofeminist views, especially those of Vandana Shiva (1989), which made an essentialist link between women and the environment and encouraged environmental programmes to focus on women's roles.

8 *Mainstreaming gender equality*   The term 'gender mainstreaming' came into widespread use with the adoption of the Platform for Action at the 1995 UN Fourth World Conference on women held in Beijing. The 189 governments represented in Beijing unanimously affirmed that the advancement of women and the achievement of equality with men are matters of fundamental human rights and therefore a prerequisite for social justice. Gender mainstreaming attempts to combine the strengths of the efficiency and empowerment approaches with the context of mainstream development. Mainstreaming gender equality tries to ensure that women's as well as men's concerns and experiences are integral to the design, implementation, monitoring and evaluation of all projects so that gender inequality is not perpetuated. It attempts to overcome the common problem of 'policy evaporation' as the implementation and impact of development projects fail to reflect policy commitments (Derbyshire 2002). It also helps to overcome the problems of male backlash against women when women-only projects are successful (Momsen 2001). In the late 1990s donor-supported development shifted away from discrete project interventions to general poverty elimination, which potentially provides an ideal context for gender mainstreaming. Attention is only just beginning to be paid to the gender dimensions of poverty alleviation (Narayan and Petesch 2002).

The Millennium Declaration signed at the United Nations Millennium Summit in 2000 sets out the United Nations' goals for the next decade (Box 1.2). These goals come from the resolutions of the various world conferences organized by the United Nations during the 1990s. Reaching these goals will not be easy but they do set standards which can be monitored (UNDP 2003). There are eight goals:

1   Halve the proportion of people living in extreme poverty between 1990 and 2015.
2   Enrol all children in primary school by 2015.
3   Empower women by eliminating gender disparities in primary and secondary education by 2005.
4   Reduce infant and child mortality rates by two-thirds between 1990 and 2015.
5   Reduce maternal mortality rates by three-quarters between 1990 and 2015.
6   Provide access to all who need reproductive health services by 2015.
7   Implement national strategies for sustainable development by 2005 so as to reverse the loss of environmental resources by 2015.
8   Develop a global partnership for development.

The principal themes

Three fundamental themes have emerged from the literature on gender and development. The first is the realization that all societies have established a clear-cut division of labour by sex, although what is considered a male or female task varies cross-culturally, implying that there is no natural and fixed gender division of labour. Second, research has shown that, in order to comprehend gender roles in production, we also need to understand gender roles within the household. The integration of women's reproductive and productive work within the private sphere of the home and in the public sphere outside must be considered if we are to appreciate the dynamics of women's role in development. The third fundamental finding is that economic development has been shown to have a differential impact on men and women and the impact on women has, with few exceptions, generally been negative. These three themes will be examined in the chapters that follow.

Women have three roles in most parts of the world: reproduction, production and community management. Today women are choosing to undertake these roles in new ways, to opt out of some, to employ paid assistance or to seek help from husbands or other family members. Planners have often used a gender roles framework but this has been criticized for ignoring political and economic differences within a community and for assuming that any new resource will be good for all women (Porter and Judd 1999). Participatory and community

Table 1.1 Regional patterns of gender differences in population dynamics, education and labour force participation rates, 1999

	Female proportion of population	Female/male life expectancy in years	Total fertility rate	Female/male HIV prevalence, % aged 15–24 years	Female/male adult illiteracy (%)	Labour force participation, female/male ratio
World Bank Group						
World	49.6	69/65	2.7	0.7/1.1	–	0.7
Low-income	49.4	60/58	3.7	1.1/2.0	48/29	0.6
Middle-income	49.5	72/67	2.2	0.5/0.6	20/5	0.7
Lower middle-income	49.3	72/67	2.1	0.2/0.2	22/9	0.8
Upper middle-income	50.5	73/66	2.4	1.5/2.2	11/9	0.6
High-income	50.4	81/75	1.7	0.3/0.1	–	0.8
World region						
Latin America & Caribbean	50.5	73/67	2.6	0.7/0.3	13/11	0.5
Middle East & North Africa	49.3	69/67	3.5	–/–	47/25	0.4
South Asia	48.5	63/62	3.4	0.3/0.5	58/34	0.5
Sub-Saharan Africa	50.5	48/46	5.3	4.5/9.2	47/31	0.7
East Asia & Pacific	48.9	71/67	2.1	0.2/0.2	22/8	0.8
Europe & Central Asia	51.9	73/64	1.6	0.4/–	5/2	0.9

Source: World Bank (2001).

development models are often gender-blind and may just reinforce local patriarchal and elite control. They often also assume a homogeneity of gender interests at the community level. To rely on such methods may well be to give official approval to the subordination of women's rights of access to a new project and to assume, unwisely, equal benefits for all community members (Momsen 2002c).

The overall framework of the book is provided by spatial patterns of gender (Seager 1997). Gender may be derived, to a greater or lesser degree, from the interaction of material culture with the biological differences between the sexes. Since gender is created by society its meaning will vary from society to society and will change over time. Yet for all societies the common denominator of gender is female subordination, although relations of power between men and women may be experienced and expressed in quite different ways in different places and at different times. Spatial variations in the construction of gender are considered at several scales of analysis, from continental patterns, through national and regional variations, to the interplay of power between men and women at the household level.

Table 1.1 provides a macro-scale view of women's position on various indicators for countries grouped according both to income level and to location. Low-income countries are characterized by populations in which women form a minority. These women bear many children and are usually anaemic while pregnant. They are poorly educated and have a low life expectancy. In most cases, as national income increases, the sex ratio becomes more balanced, life expectancy increases and women have fewer children, are healthier and better educated and participate more in the labour force.

On a continental scale, Latin America and the Caribbean have high levels of female literacy but low levels of participation by women in the formal workforce. Women in Africa, south of the Sahara, have the highest fertility rates and the lowest life expectancy, now exacerbated by the rapid spread of AIDS. In the countries of Eastern Europe and Central Asia currently in transition from socialism to capitalism, literacy rates and life expectancy are high, and so is participation in the labour force, while fertility is very low. South Asia is almost a mirror image of the transition countries as it is distinguished by the lowest proportion of women in the population and in the labour force, the lowest literacy levels and the highest levels of anaemia in pregnancy. The interrelationships between these indicators will be examined in the following chapters.

Regional trends can also be seen over time (United Nations 1995b). In Latin America and the Caribbean fertility and maternal mortality have declined but cities are growing rapidly, straining housing and infrastructure. At the secondary and tertiary levels of education girls outnumber boys, but women's labour force participation rate is lower in Latin America than in the Caribbean. Sub-Saharan Africa is the only region where the women's labour force participation rate has fallen since the 1970s, fertility is still high, literacy is low and life expectancy declined during the 1990s because of HIV/AIDS and civil strife. North Africa and West Asia have seen increased female literacy and increases in women in the labour force but both these measures are low relative to other parts of the world. In South Asia there is less gender equality in life expectancy and rates of early marriage and maternal mortality remain high.

While considering the context-specific issues of particular regions we also need to move beyond the generalized patterns of gender and development over time and space to an understanding of the realities of lives embedded in distinct localities. Broad statistical generalizations are insufficient for constructive conceptualization but the addition of oral histories and empirical field data allows us to link the local and the global through the voices of individuals. An emphasis on location and position highlights a concern with the relationships between different identities and brings a new understanding to gender and development.

Learning outcomes

- Gender roles and identities vary widely in different cultures.
- Gender equity often suffers during periods of economic stress.
- Development policies have changed over time from a focus on women only to one based on gender, sometimes including environmental aspects, and most recently to an interest in masculinities.
- On many variables there are regional similarities in the position of women relative to that of men.

Discussion questions

1   Discuss spatial variations in gender divisions of labour.
2   To what extent has economic development tended to make the lives of the majority of women in the developing world more difficult?

3   Explain why the universal validity both of gender-neutral development theory and of feminist concepts that are derived from white, Western middle-class women's experience is being questioned.

4   Why do measures describing the role and status of women display distinct regional patterns?

Further reading

Boserup, E. (1970) *Women's Role in Economic Development*, New York: St Martin's Press. This was the first book on the topic and was the stimulus for all the later work reported on here.

Cornwall, Andrea and Sarah C. White (2000) 'Introduction. Men, masculinities and development: politics, policies and practice', *IDS Bulletin* 31 (2): 1–6. Provides a review of the work done on development and masculinities.

Desai, Vandana and Robert B. Potter (eds) (2002) *The Companion to Development Studies*, London: Arnold. Contains several short articles on various aspects of gender and development by many of the leading protagonists.

Momsen, Janet H. (2001) 'Backlash: or how to snatch failure from the jaws of success in gender and development', *Progress in Development Studies* 1 (1): 51–6. Shows how a focus in development projects on women only, can lead to disaster.

Seager, Joni (1997) *The State of Women in the World Atlas*, new 2nd edition, London: Penguin Books. A very useful collection of coloured maps illustrating many aspects of gender inequality throughout the world. Includes statistics up to 1996.

Websites and e-mail

awid@awid.org (e-mail)   Address of the Association for Women's Rights in Development (AWID). Access is free to members and people with low incomes. The Resource Net Friday File comes out weekly with articles and news items on women/gender and development.

www.genderstats.worldbank.org   World Bank database with gender indicators and sex-disaggregated data for all countries in the world in five areas: basic demographic data, population dynamics, labour force structure, education and health.

www.undp.org/hdr   UNDP Human Development Report (various years).

www.un.org/depts/unsd   Women's Indicators and Statistics Database (Wistat). produced by the United Nations Statistical Division.

www.worldbank.org/gender   GenderNet website which has resources by sector and by region and links to other sources.

2 The sex ratio

Learning objectives

When you have finished reading this chapter, you should be able to:

- identify the main reasons for the differences in the proportions of men and women in national and regional populations
- appreciate changes in gender differences in life expectancy
- understand the underlying reasons for gendered patterns of migration.

It might be expected that the sex ratio, or the proportion of women and men in the population, would be roughly equal everywhere. Figure 2.1 shows that this is not so and that there is quite marked variation between countries. Explanations of these spatial patterns reveal differences both in the relative status accorded to women and men and in the quality of life they enjoy. Thus the sex ratio is often the first indication of gender inequality.

Figure 2.1 reveals that the lowest ratios of women to men are found in the oil-rich economies of the Middle East, such as the United Arab Emirates, Qatar and Kuwait, where there are high levels of male foreign workers. Other areas with low proportions of women are South Asia and China, where there is marked discrimination against women. The highest ratios of women to men are in Russia, the Baltic countries of Latvia, Estonia and Lithuania, and the transition countries of Belarus, Hungary, Moldova, Georgia and Ukraine, where male death rates are higher than those of women. The small states of Cape

Verde Islands and Djibouti also have high proportions of women, 114 and 112 respectively, per 100 men, probably because of high male out-migration (PRB 2002).

More males than females are conceived, but women tend to live longer than men for hormonal reasons. Boys are more vulnerable than girls both before and after birth. The better the conditions during gestation, the more boys are likely to survive and the more likely it is that the sex ratio at birth will be masculine. However, if basic nutrition and health care is available to the whole population, age-specific death rates favour women. In the industrial market economies these factors have resulted in ratios of about 95 to 97 males per 100 females in the general population. Sex-specific migration or warfare may distort the normal demographic pattern. Typically, however, in the absence of such factors, a female-to-male ratio significantly below 100 reflects the effects of discrimination against women. In the world as a whole there are some 20 million more men than women because of masculine sex ratios in the Middle East and North Africa, and the very marked imbalance in the huge populations of China and India. Between 1970 and 1995 the global proportion of women per 100 men fell from 99.6 to 98.6, although it increased in Latin America, South-east Asia, West Asia and Oceania but fell in Africa, the Caribbean and Europe (United Nations 1995b). The United Arab Emirates fell from 60 in 1970 to 52 women per 100 men in 2000. Many South Pacific islands had low female sex ratios of below 95 women per 100 men in both 1970 and 1995 but most showed slight improvements in the 1990s, although the Cook Islands, French Polynesia and Samoa had fewer women per 100 men in 1995 than in 1970, reflecting gender differentiated migration patterns (United Nations 1995b). In this chapter we examine the reasons for these differences.

Survival

Life expectancy at birth is the most useful single indicator of general well-being in poor countries. For the world as a whole, life expectancy is 69 for women and 65 for men and in the less developed countries 66 for women and 63 for men (PRB 2002). In the more developed world, excluding the transition countries, average female and male life expectancy at birth varies only between 84 and 77 years in Japan and 79 and 72 respectively in Portugal (ibid.), but in the developing world the range is greater, extending from 82 and

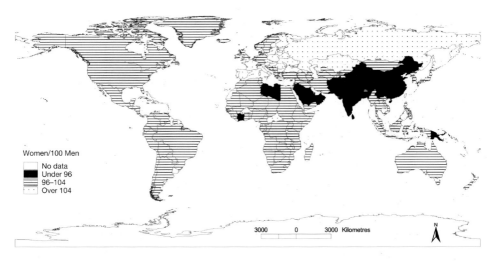

Figure 2.1 *Sex ratio, 2002.*

Sources: Sass and Ashford (2002: 4–11) and (for Iceland only) Statistics Iceland (2002: 4)

77 years for the women and men of Hong Kong to 38 and 37 in Zambia (ibid.).

Women have the shortest lives in the countries of tropical Africa and South Asia. Countries such as Burkina Faso, Tajikistan and Nepal, with similar per capita gross national incomes to those of Laos, where female life expectancy is 54 years, of approximately US$300 per year, have female life expectancies of 47, 56 and 71 years respectively (World Bank 2001; PRB 2002). These figures demonstrate that even poor countries can improve the general well-being of their women citizens by adopting a basic needs approach and ensuring that food, health care and education are accessible to all. However, within countries marked regional, class and ethnic differences may exist.

Between 1970 and 2000 life expectancy in the developing world increased, with the greatest increase being that for women (Figures 2.2 and 2.3). Women's life expectancy increased by about 20 per cent, one to two years more than the increases among men. Globally, at the beginning of the new millennium life expectancy for women averaged 69 years and that for men 65, but in low income countries the figures were 60 and 58, and in middle income countries 72 and 67 (World Bank 2001). Major explanatory factors included greater access to family planning and reproductive health care, improved nutrition and reduction in infectious and parasitic

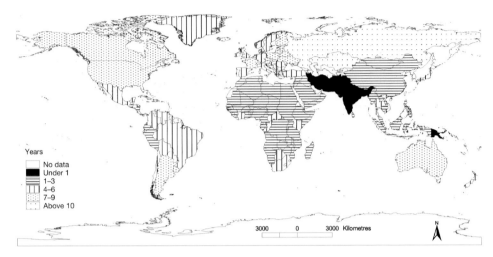

Figure 2.2 *Gender differences in life expectancy at birth, 1970–5.*
Source: United Nations (1995: 84–8)

diseases through widespread delivery of childhood vaccinations and safe drinking water. All things being equal, women live longer than men but in some countries, such as Afghanistan, discrimination against women is so severe that their average life expectancy is less than that of men. In the early 1970s men lived longer than women mainly in South Asia (Bangladesh, Bhutan, India, the Maldives, Nepal and Pakistan) and in Afghanistan, Iran and Papua New Guinea (see Figure 2.2). By the end of the century women's life expectancy was equal to or greater than that of men in all these countries except in the Maldives, Nepal and Pakistan, although there was no recent data available on Afghanistan (UNDP 2002). However, progress has been reversed in Africa. At the beginning of the new millennium the worst reversals for women were in Botswana, Lesotho, Malawi, Namibia, Zambia and Zimbabwe, where poverty and AIDS have increased and women now have a shorter life expectancy than men (UNDP 2002; World Bank 2001).

On the other hand, life expectancy for men has fallen precipitously since 1990 in many of the transition countries. This has led to the greatest gender differences in life expectancy in the Ukraine, Belarus and Russia, where women live 11, 12 and 13 years longer than men respectively (Figure 2.3).

These gender differences in life expectancy have been linked to the identification of a situation of 'missing' women. Because until

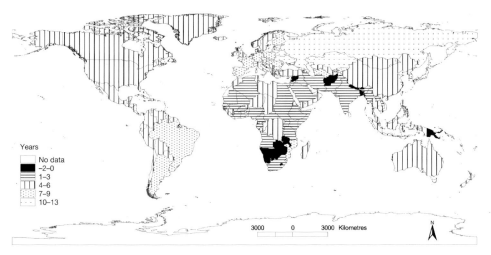

Figure 2.3 *Gender differences in life expectancy at birth, 2002.*

Sources: Sass and Ashford (2002: 4–11) and (for Iceland only) UNDP (2001: 210)

recently the greatest differences in life expectancy were found among the large populations of South Asia, China, West Asia and North Africa, it was calculated that global excess female mortality resulted in 100 million missing women (Sen 1990). More recent census data indicate that, although the absolute number of missing women has risen to between 65 and 110 million, the global sex ratio has begun to improve since 1995 (Klasen and Wink 2002). Rising female education and access to employment opportunities are associated with declines in female mortality, but this has been counterbalanced by the increased use of sex-selective induced abortion, especially in China and India, resulting in a higher sex ratio with an excess of boys at birth (ibid.).

Male and female survival chances vary at different points in their life cycle. In the first year of life boys are more vulnerable than girls to diseases of infancy and in old age women tend to live longer as they are less likely to suffer from heart disease. Any deviations from these norms indicate location and culture-specific factors. This can be illustrated by reference to sex ratios at different ages for Libya, a relatively rich country with an economy based on the export of petroleum. At every age there is a masculine sex ratio. Poor maternity care is revealed in higher death rates for women in the early and late years of childbearing when risk to the mother is greatest. This contributes to an unusual pattern of an increase in the proportion of men in the population with age. This increase is also explained by under-reporting of the female population and by the repatriation of Libyan males.

The sex ratio in South Asia

In South Asia, masculine sex ratios have become more extreme over time, with the ratio for India increasing in masculinity from 97 females per 100 males in 1901 to 94 in 2001. Spatial contrasts are very marked and have remained stable for a long period. With the exception of the small populations of the hill states, the sex ratio is most masculine in the north and the west of the region, while the south and east have more balanced or feminine ratios. Urban sex ratios are more unequal than rural, with an urban rate of 88 females per 100 males and a rural rate of 96 females per 100 males in 1995 (United Nations 1995b). The tendency for there to be fewer girls born in urban areas is related to the availability of methods of detecting the sex of the foetus and aborting those that are female. Such use of technology is now illegal in India but the relatively wealthier urban population is still able to access these methods. Masculine sex ratios are also associated with high mortality rates for young girls and for women during the childbearing years. It has been calculated that if the African sex ratio existed in India there would have been nearly 30 million more women in India than actually live today. This situation in South Asia has been linked to the general economic undervaluation and low social status of women in the region. Globally, the lowest proportions of women to men are in the rich oil-producing Muslim states of the United Arab Emirates (52), Qatar (56), Kuwait (72) and Bahrain (74), where immigrant workers are mostly male and women have low status (Figure 2.1).

In patrilineal systems, where mothers lack decision-making power, infant mortality may be high (Box 2.1). Highly stratified gender systems, where daughters are devalued, as in northern India, Korea and China, may result in high levels of mortality among girls under the age of five years. Croll (2000) indicates that son preference is both economically and culturally based in ideas of gender identity and that daughter discrimination has increased under the pressure to have smaller families. She also found that there are no clear correlations with parental characteristics and that, although college-educated mothers tend to have more daughters, more important were birth order and the gender composition of surviving children (ibid.: 26). Prenatal foeticide of female foetuses and postnatal infanticide and neglect of young girls may reduce the pressure to practise birth limitation. In India prenatal sex-determination tests were banned in 1994 but now methods to aid sex selection before conception are being advertised, which, while technically legal, are almost always aimed at avoiding the birth of a girl (Dugger 2001). Such failure to enforce laws protecting women is found in many countries as Huda (1994, 1997, 1998) has shown for divorce, child marriage, custody and inheritance laws in Bangladesh. The Indian Supreme Court in 2001 ordered the government to enforce laws against sex-determination tests and sex-selective abortions more aggressively. This decision was taken in light of the results of the 2001 Census of India, which showed that the ratio of girls to boys in the richest states of north-western India had fallen sharply over the previous decade due to the rising use of ultrasound tests to determine the sex of the foetus, resulting in sex-specific abortion. In the Punjab, for example, the ratio of girls to boys, six years old and under, has declined to 793 girls per 1,000 boys in 2001 from 875 in 1991, while in Kerala the sex ratio is in balance. The detailed regional patterns can be seen in the *Atlas of Women and Men in India*, based on 1991 census data (Raju *et al.* 1999). This situation is already having social consequences, with young men unable to find wives. In Haryana, a fairly wealthy state in north-western India, desperate fathers of sons are no longer demanding dowries from the families of eligible girls and may even offer a bride price (Lancaster 2002). At the same time the families of girls are becoming more choosy and allowing their daughters some say in the selection of husbands.

If they fall ill men are more likely than women to receive medical assistance. Illness in young girls and women is often fatalistically accepted by family members. Female infanticide has long been a tradition in many states in northern India. Indeed it has been suggested that, in some poor families, mothers feel that their daughters are better off dying as children than growing up to suffer as they themselves have. Overworked, undernourished and anaemic women tend to produce smaller babies and to be more vulnerable to the dangers of childbirth. Maternal death rates are exacerbated by the dominance of traditional medicine in obstetrics and gynaecology in many parts of the region.

Box 2.1

Female infanticide in China

In the early 1990s I was in the mountains of northern Yunnan studying rural poverty among ethnic minorities. I was a member of an international group of researchers visiting several villages. In one village, on a chilly wet day, we walked through the muddy paths and visited the school and several homes. Then most of the group, including all the men and our official translators, decided to walk to the apple orchards planted on the edge of the village. A few of us women outsiders plus one elite Yi minority woman who was at that time studying for a Ph.D. at a United States university and so could translate for us, decided to stay and talk to a village woman. For the first time we were without our official 'minders'. As we squatted around the three-stone fire in the centre of the mud-floored hut, lit only by the flickering flames from the fire, with the woman and her two young daughters we asked her about her family. She then told us that her husband wanted a son so he paid to be allowed to have a third child. When the baby was due she went to the clinic a few miles away. The baby was born safely and healthy but it was another girl. As she walked home through the fields carrying her newborn daughter, she demonstrated how she had gathered the folds of the long, thick, handwoven cotton skirt she wore and stuffed it into the mouth of her baby, suffocating her. When she got home she told her husband that the baby had been stillborn. She explained that if she had returned home with a third daughter her husband would have divorced her, blaming the sex of the baby on his wife. If that had happened, she pointed out, there would have been no one to support her other daughters, so in order to protect her older daughters she sacrificed the baby. What could we say! We held hands and mothers from three continents wept together.

Source: fieldwork, Yunnan, China, 1991.

Economic status

Urban employment opportunities for women in industry, trade and commerce are contracting and in rural areas technological change is reducing their role in agriculture, especially in the processing of crops. This decline in the economic role of women can be linked to increased discrimination against them. However, the relationship between women's role in production and the sex ratio is neither simple nor universal.

Another explanation of regional differences in the sex ratio of the Indian population is based on north–south contrasts in the transfer of property on marriage and at death. In the north, where the sex ratio is most masculine, not only are women excluded from holding property, but they also require dowries on marriage and so are costly liabilities. Sons, on the other hand, contribute to agricultural production, carry the family name and property, attract dowries into the household and take care of parents in their old age.

In the south women may inherit property and their parents may sometimes demand a brideprice from the husband's family, although dowries are becoming more common than in the past. Generally, in southern India women play a greater economic role in the family, the sex ratio is more balanced, fewer small girls die and female social status is higher than in the north. The position of women is most favourable in the south-western state of Kerala, where a traditional matriarchal society allowed women greater autonomy in marriage, and a long history of activity by Christian missionaries has helped to ensure that women are less discriminated against in access to education than elsewhere in India. The women of Kerala, with the help of women doctors, took family planning into their own hands and very quickly reduced the birth rate without government interference.

Regional patterns of sex ratios in South Asia are highly complex and vary with caste and culture. Most women have little autonomy or access to power or authority. They are faced by discrimination and exclusion and also by oppressive practices such as widow burning, known as *suttee*, which appears to be on the increase. These social constraints owe their origin to the need to protect the family lineage through the male line by controlling the supply of women. Their effect is most severe at those times in a woman's life when she is particularly physiologically vulnerable, that is below the age of five and during the childbearing years.

Sri Lanka

However, it should be noted that in one country in South Asia women do normally live four years longer than men. Sri Lanka's development process has included far-reaching social welfare programmes, especially free education and health care, for the last four decades and the benefits can clearly be seen in the improvement in life expectancy (Table 2.1). By 1967 female life expectancy, which had been two years less than that of men 20 years earlier, had surpassed male life expectancy by two years. Twenty years on Sri Lankans of both sexes have the highest life expectancy in South Asia and the additional years women may expect to live have in the last

Table 2.1 *Sri Lanka: expectation of life at birth in years*

	1920–2	1946	1953	1963	1967	1981	1987	1999	2001
Male	33	44	59	62	65	69	68	71	70
Female	31	42	58	61	67	72	73	76	74

Source: Department of Census and Statistics, Sri Lanka, for the period 1920 to 1981; World Bank (1989) for 1987 data; World Bank (2001) for 1999 data; Population Reference Bureau (2002) for 2001 data.

two decades suddenly increased from two to five, although the most recent figures suggest that life expectancy has fallen, reflecting high mortality levels in the recent civil disturbances. Yet Sri Lanka still has a masculine sex ratio, with only 49.2 per cent of the population being female in 1999, despite the huge improvement in female life expectancy. It is also one of the few countries where chronic malnutrition is worse for girls under five years of age than for boys (United Nations 1995b). Clearly patterns of sex ratios and life expectancy are complex and unstable.

Migration

Sex-specific migration also affects sex ratios (Figure 2.4). In Libya, during the 1970s and 1980s, a booming economy suffering from a labour shortage attracted many foreign workers and by 1983 these foreigners made up 48 per cent of the workforce. About three-quarters of the foreign residents were male because the Libyan Government perceived men as most suitable for the type of work and the living conditions available. Thus the overall sex ratio of Libya in 1985, even after declining fortunes in the oil industry had led to the departure of many foreign workers, was 111.4 males per 100 females, compared to a ratio of 104.2 per 100 for the citizen population. In 2000 the gender ratio among foreign-born residents in Libya was 227 men to 100 women, indicating a continuing dominance of single male migrants (United Nations 2000).

Migration is a phenomenon associated with spatial differences in employment opportunities. Migrant workers, worldwide, come predominantly from countries which cannot find jobs for all their workforce at home. Examples of such 'labour reserves' are Botswana and Lesotho in southern Africa, and the West Indies. These areas have feminine sex ratios, with 91 men per 100 women recorded for

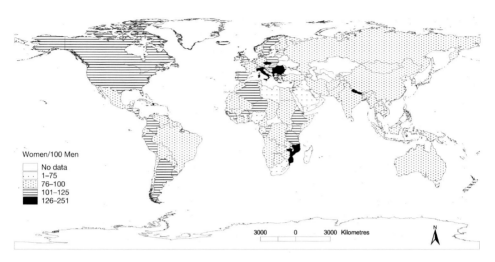

Figure 2.4 *Sex ratio of international migrants, 1990.*

Source: United Nations (2000: 17–21)

Botswana, and 93 for Lesotho (see Box 2.2) and Montserrat, a
British colony in the Caribbean where men have migrated out for
decades. Among in-migrants the sex ratio is mostly masculine in
Yemen, Sierra Leone, Qatar, Bahrain and Lebanon, while countries
with more women in-migrants, including refugees, are Nepal, the
Czech Republic, Romania, Mozambique, Haiti, the Balkans and Italy
(Figure 2.4).

Many people left the tiny Caribbean island of Montserrat in the
1950s and 1960s to work in Britain. The 1960 census recorded only
78 men for every 100 women. For the age cohort over 70 years there
were fewer than 40 men per 100 women, although the sex ratio was
masculine for the under-15s. Thus Montserrat society became
predominantly one of grandmothers and children, with very few men
of working age left behind on the island. After 1962 migration
became more difficult because of legal barriers introduced by the
governments of the main receiving countries. Gradually Montserrat's
prosperity improved as foreign residents and businesses were
attracted by the stability offered by the island's colonial status. Many
former migrants, having either reached retirement age or lost jobs
because of recession overseas, decided to return to the land of their
birth, and the island's population began to increase after a long
period of decline.

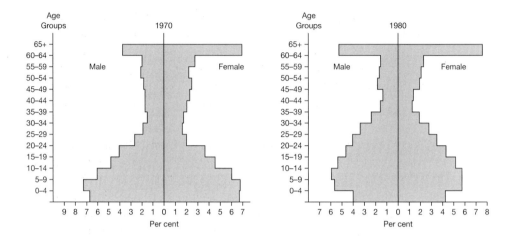

Figure 2.5 *Montserrat, West Indies: age and sex structure, 1970 and 1980.*

Figure 2.5 shows the narrow-waisted population pyramid produced by these fluctuations in migration patterns. Birth rates were affected by the absence of people of reproductive age and fell from 29.5 per 1,000 people in 1960 to a low of 17.7 in 1976 and then recovered to 22.3 in 1982. Mortality rates in the first year of life fell from 114.2 per 1,000 live births in 1960 to only 7.7 in 1982. The island still had a high proportion of elderly females, now added to by return migrants, often as pensioners, and foreign retirees. Severe hurricanes in the late 1980s discouraged further return migration and caused considerable damage but the eruption of the Soufriere Hills volcano in 1995, after almost four centuries of dormancy, was the final blow. Many islanders fled as refugees to neighbouring islands and to Britain. The foreign second-home owners left, never to return. By late 2002, as signs of a new eruption became more persistent, the population of the island had fallen from 11,000 to 4,000.

Montserrat had a sex ratio of 94.9 men per 100 women in 1988, the most masculine ratio recorded in any census since 1871, when it was 84.4, falling as low as 72.5 in 1921. The long history of female numerical dominance on this island has contributed to women's economic importance and independence. In 1972 women operated 44 per cent of small farms on Montserrat, but this had fallen to only 23 per cent in 1983, as male return migrants replaced female farmers and women took advantage of better-paid employment elsewhere in the economy. At the same time, women managed to retain their dominance of the prestigious jobs in the civil service and local financial sector into which they had moved during the period of mainly male out-migration. Montserratian men explain this by relying on the now fallacious argument that there are more women than men of working age on the island. Women were also able to continue to take advantage of the universal, free childcare which the government had been forced to introduce when there were few men available for the workforce.

Both men and women migrate but the reasons for the migration, the type of destination and the length of time spent at the destination are often gender specific. In so far as any general patterns can be identified, men are more likely than women to migrate in order to gain educational qualifications, while women are more likely to migrate to marry or to rejoin a migrant spouse, but autonomous female migration is increasing in importance, especially among younger women (see Box 2.2). Migrant women may also be flouting traditional patriarchal restrictions and norms. They may be avoiding arranged marriages, leaving a marriage that is unhappy or has not produced children, or escaping from low economic and social status. In the transition countries, international migration has only become legally possible since 1989 and much of the current movement involves trafficking in women for sexual purposes from the poorest parts of the region, such as the Ukraine, to Western Europe.

Migration for both men and women may be short-term or circular rather than permanent and this temporal pattern will affect both the source region as well as the adaptation of the migrant to the receiving area. Remittances to family left behind are most consistent from transnational migrants intending to return, and regular visits by migrants bring new ideas into traditional rural areas. Teenage Indian women from the highlands of Peru are often sent to the cities to work as servants but are expected to return to their villages to marry. In Indonesia both men and women move between rural and urban areas in a circular manner, responding to gender-specific labour demands in the countryside during the agricultural year.

Rural-to-urban migration involves the largest number of people but movement may also be from rural to rural areas or across international boundaries (see Plates 2.1 and 2.2). Three factors affect female rural to urban mobility: female participation in agriculture, availability of economic opportunities for women in the cities and socio-cultural restrictions on the independent mobility of women. Internal migration from rural to urban areas is dominated by women

Box 2.2

Sex-specific migration and its effects on Lesotho

For many years, the women of Lesotho were gold widows: left behind to make of life
what they could, while the men worked in the mines of South Africa. The border
between the two countries formed a sex-specific barrier from the 1960s until the mid
1990s, with only men allowed to enter South Africa to work. Lesotho society was,
therefore, one in which women did most of the work of agriculture and social
reproduction. Yet women were generally better educated than men and had developed
considerable autonomy but had no power within the family. This led to a deep sense of
frustration among women because they were denied access to the modern industrial
world. All that most of them could do was remain on South Africa's periphery,
reproducing its labour force, doing unpaid domestic work, cultivating infertile soil,
seeking low-paid local employment, providing a market for South Africa's goods and
becoming increasingly dependent on the unreliable supply of remittances from male
wage earners.

Traditionally women moved to their husband's home on marriage, so they have
dominated the pattern of rural-to-rural moves while men have played the major role in
international migration. Now women have started to move to the town and, in a survey
of young urban migrants in 1978, it was found that the sex ratio of migrants under the
age of 24 years was 49 males per 100 females (Table 2.2). Women were moving because
life in the rural areas had become intolerable. About two-thirds of rural households were
headed by women but only about half of these had migrant husbands and received
remittances; the rest were widows, reflecting the high rate of fatal accidents in South
African mines. Soil erosion and declining fertility has made it increasingly difficult to
live in rural areas, where marijuana is said to be the main crop, so urbanization is
increasing rapidly (Abildtrup and Rathmann 2002).

Table 2.2 *Maseru City, Lesotho: age of migrants at time of move, by sex, 1978*

	Age				
	Under 24	25–34	35–44	44+	Percentage of total
Percentage of men	32.8	32.8	23.0	11.5	45.3
Percentage of women	55.7	24.9	12.2	7.2	54.7

The well-educated young women of Lesotho are faced with a situation in which South
African 'influx control measures' keep them penned within their impoverished little
country, where they are further controlled by a traditional, patriarchal society. They see
Maseru, the capital city, as providing them with some hope of freedom and, in 1978,

43 per cent of female migrants declared that they would never return to their rural homes, compared with only 30 per cent of the men. Yet employment is often illusory. A survey of de facto household heads in the city showed that 53 per cent of the women but only 17 per cent of the men were unwaged. Consequently large numbers of women were forced to take up petty trading, scavenging, the illegal sale of home-brewed beer and prostitution. Where jobs in the formal economy were obtained by women, they were predominantly in the low-paid clerical and domestic service sectors (Table 2.3).

Table 2.3 Occupations of migrants to Maseru City, Lesotho, 1978 (%; N = 416)

Occupations	Men	Women
Labouring	9.0	0.0
Construction	21.8	0.4
Engineering and driving	13.3	0.4
Clerical	10.1	13.2
Sales	6.9	3.5
Professional and managerial	14.3	7.0
Domestic	1.6	19.7
Others	7.5	4.4
Unemployed	15.4	51.3

Prostitution is one of the better-paid jobs. In its modern form it is associated with the growth along the border of Las Vegas-style entertainment centres catering for white males from South Africa. Weekend trips for this purpose became part of Lesotho's tourist trade. Thus the border is permeable in both directions by only one sex. Lesotho men served South Africa in her mines while Lesotho women were forced to stay at home and educate themselves to provide services for white South African males, services which were illegal within apartheid South Africa.

In post-apartheid South Africa the economy of Lesotho has changed. In 1996 there were more than 100,000 men from Lesotho working in South Africa's mines but by 2000 it had fallen to 60,000. This is reflected in the change in the ratio of miner's remittances to GDP from 60 per cent in the late 1980s to 26 per cent in 1999 (Abildtrup and Rathmann 2002).

Lesotho remains among the countries defined as having medium human development but, after steady improvement on the Human Development Index from 1975 to 1990, it has been one of the few countries to show a steady decline in its score on the HDI in 1995 and 2000, with two-thirds of the population living on less than US$2 a day despite a low adult illiteracy rate of only 17 per cent (UNDP 2002). A new aspect of the negative influence of male return migration and the dependence of many Lesotho women on prostitution is seen in the HIV/AIDS infection rates: in 2001, 31 per cent of adults aged 15 to 49 were living with the disease, of whom 55 per cent were women (Figure 4.3). Although 93.6 per cent of women were literate compared to only 72.5 per cent of men, the life expectancy of women had fallen below that of men in 2000 to only 45.6 years, largely because of HIV/AIDS. Thus Lesotho joined a group of only eight countries worldwide where women's life expectancy was lower than that of men (UNDP 2002). Local customs also encourage the spread of AIDS. A husband returns from the mines infected, his wife has to care for him until he dies, then his younger brother takes over his wife and so the disease spreads. Another cultural factor quoted by Abildtrup and

Rathmann is that 'when a woman is breastfeeding, she is not supposed to have sex for at least two years. During that period the wife or the man's mother look around for a decent woman who will be able to provide comfort for the man' (2002: 84). That decent woman is someone else's wife, as according to Lesotho custom single women do not have sex. The arrangement is a secret between the two women and neither of the husbands involved has any decision-making role. In addition, women's subservience to their husbands means that they have little control over taking preventive measures against AIDS.

An army rebellion was put down with the help of South African military intervention in 1998 and the new democratic government of Lesotho is increasingly seeking foreign domestic investment that can provide employment mainly for women. The economic activity rate of women, despite their education, is still low at 47 per cent in 2000, which was only 56 per cent of the male economic activity rate (UNDP 2002). In 2001 Lesotho had a total unemployment rate of 45 per cent but the government offered tax concessions especially for foreign companies locating outside Maseru and wage rates for trained workers in the textile industry were below those of Asia (Lamont 2001). Foreign investors are mainly from Taiwan, seeking to take advantage of the African Growth and Opportunities Act (AGOA) signed in 2000, which allows duty- and quota-free access to the US market for textiles and clothing. In 2001 these exports provided 29 per cent of the nation's GDP (Abildtrup and Rathmann 2002). As a least developed country Lesotho also has preferential trade access to the European Union. Employment in the textile sector rose from 19,000 to about 40,000 following the signing of AGOA but the agreement expires in 2008. Lesotho had attracted 19 Asian companies, mainly in the textile and footwear sectors by 2001, but working conditions are harsh and the factories are heavy polluters (ibid.). To what extent new employment opportunities at home will affect migration to South Africa, and the extent of the AIDS epidemic impact on the costs of industry, has yet to be seen.

Sources: Abildtrup and Rathmann (2002); Lamont (2001: 7); Wilkinson (1987); UNDP (2002).

in Latin America and parts of South-east Asia and by men in Africa, South Asia and the Middle East, reflecting regional differences in the gender-specific pattern of labour demand. As a consequence of migration, the sex ratio for Latin American cities in the period 1965–75 was 109 women for every 100 men. Women continue to move to the cities and by 2000 a higher proportion of women than men were in urban areas in all countries of Latin America (ECLAC 2002b). On the other hand, in African towns the ratio was 92 women per 100 men in the 1970s. In colonial Africa women were discouraged from migrating to the towns and in Uganda in the 1950s all single women in Kampala were considered to be prostitutes and were by law repatriated to the countryside. Today rural poverty

Plate 2.1 *Brazil: migration to the colonization frontier. A family made up of a sister and brother from Japan had been attracted by the opportunity for land ownership in a colony in Maranhão. The man is standing beside his seed bed and the well provided by the colonization agency. The traditional gender division of labour continues, with the woman still doing most household tasks, despite migration and an unusual family structure.*
Source: author

and backbreaking farmwork is driving women to the cities, where they can find opportunities for education and economic independence.

International and transnational migration

Recent theoretical work on international migration has argued that the term 'transnational migration' better reflects the reality of migrant experience during the twentieth century (Boyle 2002). There is debate as to how new this is and how to define it, with it being seen as a mode of cultural reproduction, as a type of consciousness and as a reconstruction of place. Transnational migrants actively maintain simultaneous social and economic relations linking their place of origin and destination. By maintaining these links migrant groups deterritorialize nation states and so conflate the social and the spatial. Itzigsohn and Saucedo (2002) found that the level of incorporating in the receiving country does not affect transnational behaviour but

Plate 2.2 *Fiji: a family of new settlers in interior of the island. They were in their second season on the land, had built a small thatched hut to live in and were still clearing forest.*
Source: author

there are different explanations for this behaviour for different national migrant groups. Women appear to participate more in transnational activities related to household management, while men operate in the public spaces of socio-cultural and political transnational links (Itzigsohn and Saucedo 2002). Women migrants were often seen as 'trailing spouses' but there is increasing feminization of international migration. Men and women experience migration differently and this contrast affects patterns of settlement and return (Box 2.2). Migration itself also impacts gender relations, with many migrant women becoming more independent and resistant to patriarchal pressures. This new independence, not entirely based on their earning power, may make women less likely than men to return to their natal countries. Migrant women are often involved in transnational motherhood and have to find ways to fulfil their role as mother to children who live in different countries, at the same time as they act as surrogate mothers to other women's children (Momsen 1999). Both mistresses and maids are trapped in the binds of domesticity and both suffer from the guilt of not spending time with their own children.

Most variations in male and female spatial mobility relate to the relative opportunity costs of moving. Men's greater mobility, to more places, over a broader age span and on a more independent basis, reflects their relative detachment from reproductive activities in the household (Chant 1992). The same negative link between mobility and reproductive duties is seen in the propensity of women migrants to be young and single. Men tend to migrate over longer distances and to participate in international migration more than women. Males exceeded females in 83 per cent of all annual international migration flows between 1967 and 1976, with the exception of movement to the United States.

Since 1930 women have constituted a majority of the foreign-born legal immigrants to the USA (61 per cent of those admitted between 1952 and 1978), and Asia, Latin America and the Caribbean have replaced Canada and Western Europe as the leading sources of immigrants. The Caribbean, the Philippines, Thailand and Turkey provide significant flows of autonomous women migrants across international boundaries. By 1990 the proportion of women in international migrant flows had increased (Figure 2.5), with over a third of countries, 64 out of the 182 for which data was available, having a majority of women immigrants. Female migrants make up an especially high proportion from the populations of Eastern Europe, the Caribbean and the Philippines. On the whole, women migrants are more likely to keep in touch with their home community and to send money and goods back, especially to their children and to their own mothers. These migrants are becoming an increasingly important part of Appadurai's (1996) global 'ethnoscape', made up of tourists, refugees, exiles, guest workers and immigrants who are increasingly affecting the politics of and between nations.

Men and women migrants compete in separate labour markets. Among the vast majority of migrants who are poor and unskilled, men find a great variety of job opportunities available but women migrants tend to be concentrated in poorly-paid service sector jobs (see Box 2.2). Government policy in receiving countries causes changes in the sex ratios of migrants over time, depending on whether it is aimed at worker recruitment (and what type of work) or at family reunification.

The deskilling of industry in the developed world has created a demand for people willing to work long hours at boring, monotonous

jobs for low wages and so provided a niche for immigrants. Many of
these immigrants enter on restricted permits linked to work for one
employer or to the legal status of their spouse. If their marriage
breaks down or they lose their original job they may be deported.
In this situation, immigrants become the most exploited workers.
They cannot complain if they have to work for less than the
minimum wage and for very long hours. They often find jobs in
hotels, restaurants, nursing homes, domestic service or food
processing.

Another opportunity has arisen as a consequence of the increased
proportion of employed married women in the richer countries
which has expanded demand for domestic servants. Thus Canada
supported a programme to bring in West Indian women to work as
domestics, Sri Lankan women are sent to the Middle East, and
migration from the Philippines to the United States, Canada,
Western Europe, Hong Kong and Singapore is female dominated,
reflecting active recruitment of domestic workers (Momsen 1999).
In the mid 1970s 57 per cent of all long-term work permits issued
by Britain to Filipinas were for domestic work, but many women
working as domestics are illegal, undocumented migrants and so
very vulnerable to abuse from their employers (Momsen 1999;
Anderson 2000). In the early 1990s some 1.7 million female
domestic workers were thought to have left their homes in the
Philippines, Sri Lanka, Indonesia and Bangladesh to work
elsewhere in Asia and the Middle East (Yeoh and Huang 1999).
Domestics remitted around US$75 million to Asian countries in
1995 (Momsen 1999). There is also a racial and religious hierarchy
of domestic workers, with Filipinas and East Europeans replacing
West Indians in Canada, and Filipinas being preferred to Africans
in most of Europe, while Muslim Sri Lankans, Indonesians and
Bangladeshi women are preferred in the Middle East (ibid.).
The economic importance of these migrants has, in most cases,
prevented their home governments from enforcing regulations to
improve the conditions under which they are hired.

Effects of migration on rural areas

When men migrate, leaving their wives and families behind in rural
areas, the rural economy is affected. Men often continue to exert
control over household finances and decision-making in Africa,
Costa Rica and the Caribbean (Chant 1992). In Kenya and Zimbabwe

two-fifths of rural families are headed by women and these women have a heavy burden of work, leaving little time for leisure. Migrant husbands' decision-making authority in their household or native village leads to delays in the implementation of community projects and to a situation in which wives, who are expected to look after the cattle, may not sell or slaughter a beast without their husband's permission. The men have little incentive to use the land more efficiently as most of their income comes from the town. They see the land as a cheap place in which to raise their children and somewhere to retire. But, on the other hand, wages sent home by men working in the cities enable many of these families to survive the hard times. Research in Zimbabwe showed that households receiving cash from a migrant earned a third more from farming than those without remittances because they were able to buy modern inputs such as fertilizer.

When African women migrate independently to urban areas they are breaking the patriarchal controls and failing in the heavy productive and reproductive duties they have to cope with in rural areas. Such behaviour often results in social penalties and criticism (Nelson 1992). Husbands do not want their wives in town because they lose the benefits of their rural productivity, and there are few respectable jobs for uneducated women in urban areas. Nelson (1992) suggests that this leads to an increased burden of work for husbands and boredom for wives. Illegal activities, such as brewing beer and prostitution, brought considerable financial rewards for women until the 1980s, when increased police pressure, large-scale commercial production of beer and the spread of HIV/AIDS made these activities less lucrative and more risky.

Without remittances, households headed by single mothers in rural areas are significantly poorer than male-headed households (Box 2.2). Men farmers in Botswana in the 1970s were twice as likely to own the cattle needed for ploughing, milk and financial security than women farmers and their crop yield was generally four times greater. In colonization zones of eastern Colombia wives abandoned by migrant husbands found it very difficult to break with tradition and take on the role of farmer even though their children were hungry (Townsend and Wilson d'Acosta 1987). In the Caribbean, despite a long history of women farmers, male migration also causes problems. Women's shortage of time and difficulties in obtaining assistance with farm tasks considered to be male, especially pesticide application, have led to a decline in agricultural output and

Table 2.4 *Gender differences in migration on small-scale farms in the eastern Caribbean*

	Barbados (N = 128)		St Lucia (N = 68)		Nevis (N = 99)		Montserrat (N = 66)	
	Men (%)	Women (%)	Men (%)	Women (%)	Men (%)	Women (%)	Men (%)	Women (%)
Total farmers	73	27	84	16	68	32	67	33
Returned migrants	43	7	47	20	67	28	46	9
Remittances: none received	94	83	28	55	39	22	50	45
Remittances providing over half of income	4	14	n/a	n/a	16	44	0	23

Sources: Author's field surveys: Barbados, 1987; St Lucia, 1971; Nevis, 1979; Montserrat, 1973.

underutilization of land. Thus the feminization of agriculture is often accompanied by increased poverty and sometimes by malnutrition in rural families. More West Indian women than men farmers are dependent on remittances from migrant relatives (Momsen 1986) but these funds are seldom invested in agriculture (Table 2.4). Instead they are used to improve rural living conditions, to finance migration for other members of the family and even to allow women to retire from the hard labour of farming (Momsen 1992a). Men are more likely to migrate when young, as women are more spatially restricted by childcare, although children are increasingly left in the care of grandmothers. Men may return to farm in retirement but are less likely to be innovative or highly productive farmers than when younger. As women stay home to raise their children they are more likely to receive remittances from these children when they migrate, if not from husbands, whereas fathers have often lost touch with their children. However, the person with the highest dependence on money sent from overseas in Barbados (Table 2.4) was a young man who received 75 per cent of his income from his mother working in Britain. In societies where migration, whether repeat, circular or return, is a common pattern, households are extended to include those relatives living overseas. In these cases the household is not so much a residential unit but a group of people engaged in the pooling of goods and services.

Such an extension of the concept of household is reinforced today by the use of electronic communications both between migrants and their natal home and between migrant communities in different parts of the global diaspora. Many migrants see themselves as living transnational lives, while returning home regularly for family and local festivities. Such frequent contact combined with dedicated websites allows migrant communities to contribute to the development of their natal island, as in the case of the Caribbean island of Carriacou, where migrants, currently living in New York and Toronto, have provided equipment for the island's hospital (Mills 2002). Such links become vital at times of crisis, as in Montserrat following the 1989 hurricane and the volcanic eruptions of the mid 1990s.

Female-headed households

Labour reserves export their excess male labour and are left with a society made up of families headed by women. In the Caribbean about one-third of household heads are women. This proportion ranges from 50 per cent in St Kitts and 44 per cent in Montserrat to less than 20 per cent in Guyana. In Brazil, although the total for the country as a whole is only around 15 per cent, spatial variation is also marked: female-headed households are most common in the very migration-prone, arid north-east and in urban rather than rural areas. For the developing countries as a whole, it is estimated that about one-sixth of all households are headed by women, with the highest regional figures being found in southern Africa (43.3 per cent) and the Caribbean (34.3 per cent), while the lowest are in South Asia (9.1 per cent) (Varley 2002). The transition countries of Eastern Europe have a regional figure of 26.9 per cent (ibid.). Many people believe that the proportion of female-headed households is increasing rapidly as a result of modernization and globalization but the evidence is not clear cut and in some countries it appears to be declining (Momsen 2002a). It has been suggested that, by seeing female-headed households only in negative terms and linking them with poverty, publicizing an increase may justify gender and development (GAD) policies (Varley 2002).

Female-headed households may be the result of the breakdown of male-headed households through death, marital instability or migration. They may also occur in a situation where the woman has no permanent partner or when the husband has several wives. Regional patterns are distinguishable: in Asia widowhood is still a prime cause; in southern and North Africa and the Middle East international migration is the predominant reason; in West and

Central Africa male migration to cities leaves women alone in rural areas; and in the Caribbean many women choose to have short-term visiting relationships but no permanent resident partner and this pattern is exacerbated by international migration.

In societies where property is corporately held and the household is the unit of labour, women rarely emerge as heads of households. Female-headed households will develop where women have independent access to subsistence opportunities through work, inheritance, or state-provided welfare and are permitted to control property and have a separate residence. Their subsistence opportunities must be reconcilable with childcare and must provide an income not markedly lower than that of men of the same class. Development has been accompanied by increased privatization of the means of production and a decline in cooperation within kin groups and has thus provided the conditions for the growth of female-headed households.

These households are often among the poorest as they contain fewer working adults than male-headed households and women earn lower wages than men. Their composition has also been said to constitute a poverty trap, with children disadvantaged because they may have to leave school early to seek paid employment or take over household chores to allow the mother to work outside the home. But it has also been shown that single mothers are more likely to send their daughters as well as sons to school and to invest in their children (Momsen 2002a). Maternal neglect and lack of paternal discipline has been thought to encourage truancy and delinquency and to perpetuate a familial pattern of deprivation. However, households headed by women are not undifferentiated and should not necessarily be seen as victims of development. In some cases women choose to establish their own household in order to gain decision-making independence and to escape male violence and economic reliance on an irresponsible man. Such households have a positive effect on female autonomy and, despite suffering from stigmatization as a deviant form, many function very successfully both socially and economically.

Learning outcomes

- Uneven sex ratios are the result of discrimination against women and gender-selective migration.

- Gendered patterns of life expectancy are unstable and, after many decades of steady increase, the last decade has seen declines in Africa and among men in the post-communist countries.
- Migration is now being undertaken by both men and women independently and many migrants lead transnational lives, keeping their connections with their homeland while they work elsewhere.
- Female-headed households are not necessarily the poorest.

Discussion questions

1 What are the main reasons for high female sex ratios and high male sex ratios?

2 Describe the different types of female-headed households and relate these differences to the likelihood that these households will be poor.

3 Why is independent migration by women from poor to rich countries increasing?

4 How is new technology affecting both sex ratios and migrant remittances?

Further reading

Chant, Sylvia (ed.) (1992) *Gender and Migration in Developing Countries*, London: Belhaven Press. An edited collection of case studies of gendered migration in Costa Rica, Peru, the Caribbean, Ghana, Kenya, Bangladesh and Thailand. The introduction and conclusion offer a theoretical framework and discussion of the policy implications of gender-selective migration for development.

Croll, Elizabeth (2000) *Endangered Daughters: Discrimination and Development in Asia*, London: Routledge. A study of discrimination against women, looking mainly at India and China.

Momsen, Janet H. (ed.) (1999) *Gender, Migration and Domestic Service*, London and New York: Routledge. Contains case studies from Africa, Asia and Latin America, with a review chapter by the editor identifying common patterns in terms of migration for domestic service.

Momsen Janet H. (2002) 'Myth or math: the waxing and waning of the female-headed household', *Progress in Development Studies* 2 (2): 145–51. Discusses the various types of female-headed households and examines the myth that they are always among the poorest.

Sweetman, Caroline (ed.) (1998) *Gender and Migration*, Oxford: Oxfam GB. Originally published as an issue of the journal *Gender and Development*, it includes eight articles covering trafficking, seasonal migration, migration for domestic work and the problems experienced by migrants from developing countries in adjusting to life in industrialized countries. The articles emphasize the importance of both economic and social factors in the decision to migrate, while also showing that the action of migrating has a social as well as an economic outcome.

Websites

www.eclac.ch/publicaciones/DesarrolloSocial/3/LCG2183PI/PSI_2002_Summary.pdf Economic Commission for Latin America and the Caribbean (ECLAC) (2002) *Social Panorama of Latin America 2001–2002.* Briefing Paper that provides detailed current statistics on Latin America and the Caribbean. The tables, in most cases, offer gender, age and urban/rural breakdowns on population, education, economic and employment information for the individual countries of the region.

www.iom.ch   The International Organization for Migration (IOM) acts as an electronic clearing house for migration-related information.

www.iom.ch/migrant_rights   The International Organization for Migration also has a website dedicated to migrants' rights which is intended to share information on migrants' rights and to provide linkages to other sources.

3 Reproduction

Learning objectives

When you have finished reading this chapter, you should be able to understand:

- the meaning of social and biological reproduction
- the importance of gender differences in education
- women's use of time
- state interventions and control of women's bodies for population planning.

The term 'reproduction' is a chaotic concept which not only refers to biological reproduction but also includes the social reproduction of the family. Biological reproduction encompasses childbearing and early nurturing of infants, which only women are physiologically capable of performing. By social reproduction is meant the care and maintenance of the household. This involves a wide range of tasks related to housework, food preparation and care for the sick, which are usually more time-consuming in developing countries than in the industrialized world. In most countries women are also expected to ensure the reproduction of the labour force by assuming responsibility for the health, education and socialization of children. Poor countries generally offer less state assistance for these tasks than is provided in post-industrial countries.

In addition to household maintenance, social reproduction also includes social management. This latter role of women is often ignored. It involves maintaining kinship linkages, developing neighbourhood networks and carrying out religious, ceremonial and social obligations in the community. The survival strategies of many poor women depend on their success in this role. Local and kin groups can help when members of the family become ill, need a job or a loan or are faced with some other sort of crisis. A woman's success as a social manager may bring status to her family and to herself and enable her to take on leadership positions within the community.

Reproduction may be distinguished from production on the basis of the law of value. Reproductive labour has use-value and furnishes family subsistence needs, while productive labour generates exchange-values, usually cash income. Empirically this separation is very difficult to make as, within the domestic sphere in which most women work, both categories of tasks are interrelated and enmeshed in a totality of female chores. Any one task may have both use- and exchange-value at different points in time. Yet it is analytically useful to accept this division as a theoretical framework within which to consider the diversity of women's domestic labour.

It is increasingly being realized that the task of reproduction is a major determinant of women's position in the labour market, the gender division of labour and the subordination of women. The household is the locus of reproduction so that social relations within the household play a crucial role in determining women's role in economic development.

With modernization and industrialization, unpaid housework becomes increasingly isolated and spatially separated from paid productive work outside the home. Women's participation in the productive labour force will inevitably be affected by the time and energy burden of their reproductive tasks as well as by the power relationships between household members. Large families can be seen as an opportunity cost for women, limiting life choices or ensuring support in her old age. In order to understand fully the nature of women's subordination and their role in the development process, it is essential to study both reproduction and production and the interrelations between them.

Friedrich Engels (1820–95), a close associate of Karl Marx, saw reproduction as the key to the origin of women's subordination by men. He believed that it was associated with the introduction of the concept of private property. The wish of the property owner to pass his property on to his children led to the need to identify the paternity of these heirs by controlling women's sexuality, and then to ensure their survival by regulating her reproductive activities. However, Engels assumed that woman's participation in productive activities, as a result of the spread of industrialization, was a necessary precondition for her emancipation. It is now clear that women's increasing involvement in the wage economy in the developing world has not ended their subordination. Rather, it has been accompanied by the transfer of patriarchal attitudes from the household to the factory, and the desire to seclude women within the family has encouraged outworking in the home at very low wages. Development has not always brought greater freedom for women and in many cases women are now expected to carry the double burden of both reproductive and productive tasks.

Women in the Middle East and North Africa have the lowest rates of economic activity (Table 1.1) and this trait is linked to Islamic patterns of seclusion of women, facilitated by national wealth which allows the state to use foreign migrants to replace the need for citizen women workers. Marriage is compulsory for the faithful but is legally an unequal institution: men may have up to four wives and infertility or failure to bear sons are grounds for divorce. In many Muslim countries women are not only segregated from men but have seclusion or *purdah* imposed on them and have to wear long, concealing garments and sometimes a veil in public.

Biological reproduction

Fertility, as measured by the total number of children born, on average, to each woman during her reproductive years, is probably the best documented aspect of women's lives, for reasons which would have been clear to Engels (Figure 3.1). On the whole, fertility rates in the South are much higher than elsewhere. At any one time it is thought that one-third of women in developing countries are either pregnant or lactating, although fertility rates are declining. In The Gambia the average woman has 10 to 14 complete pregnancies and spends virtually all her reproductive years either carrying a child in her womb or breastfeeding a baby. The physiological stress of this reproductive activity and its effect on the woman's ability to undertake tasks related to household maintenance, such as collection of water, fuel gathering, food processing and subsistence farming, need to be considered in development planning.

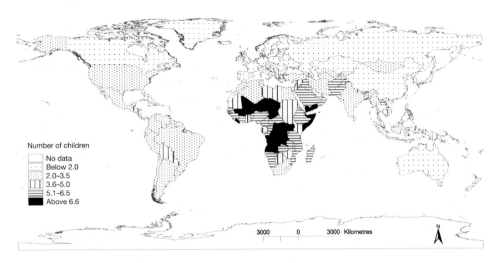

Figure 3.1 *Total fertility rate, 2002.*

Sources: Sass and Ashford (2002: 4–11) and (for Iceland only) Statistics Iceland (2002: 5)

Fertility rates vary enormously from country to country (Figure 3.1). At the beginning of the third millennium those countries with total fertility rates of six or more children were all but one, Afghanistan, in sub-Saharan Africa. In these countries male migration to cities or involvement in warfare left women dependent on their children's labour for help in subsistence farming. The lowest rate in Africa, 2.9 children per woman, is in the richest country with the most diversified economy, South Africa. Fertility rates below replacement level, typical of post-industrial societies, are found in the former communist countries, in the city states of Hong Kong and Singapore, in the Caribbean island states of Barbados, Cuba and Trinidad and Tobago, and in Thailand. These are all countries with high levels of female education and easy access to family planning. In Latin America there is considerable variation from levels of 2.3 children per woman in Chile and Uruguay to 4.8 in Guatemala, reflecting differences in education levels and attitudes to family planning (PRB 2002). Those countries with high fertility rates tend to have the large numbers of poor illiterate women often associated with a high proportion of rural, indigenous people. Clearly fertility rates are related to levels of development, although it would be wrong to assume that large families are always considered negatively by women. As women move to cities, become better educated and find new opportunities for work and self-development outside the home, the birth rate tends to fall. In cities children are less useful as supplemental labour and are more costly to maintain.

Fertility has been declining at an accelerating pace throughout the world. In the early 1970s about 60 per cent of all countries had a total fertility rate of 4.5 births per woman or higher. By the first half of the 1990s, this proportion had fallen to 40 per cent and one third of countries had a total fertility rate of fewer than 2.5 births per woman (Bulatao 2000). Eight explanations have been advanced for this fertility transition:

- mortality reduction of infants and children;
- higher costs of raising children and reduced economic contributions from children as years in school increase;
- opportunity costs of childbearing for parents, especially mothers;
- transition from extended to nuclear families leading to changing values and gender roles;
- traditional societal support for large families declines with modernization;
- improved access to contraception and abortion;
- later marriage;
- increased spread of ideas and practices which encourage lower fertility.

All of these explanations have a gender dimension. Increased access to education for women leads to later marriage, the ability to raise healthier and better-fed children so reducing infant and child mortality, knowledge of family planning, higher opportunity costs of children as women are able to obtain better paid jobs and recognition of the value of education for both boys and girls. Cultural norms may still lead to child marriages as in Bangladesh, where 5 per cent of 10–14-year-olds were married in 1996, and in northern India, Nigeria and Ethiopia even earlier marriages are not uncommon (Crossette 2001). However, laws requiring that all marriages be registered and that both partners demonstrate their consent have resulted in the average age at marriage for women in Sri Lanka rising rapidly from the late teens to 25 years of age today (ibid.). Delaying the age of marriage directly affects the number of children women bear.

In addition to the link between a woman's education level and her decision as to the number of children she wants, more general changes in society also affect fertility rates. Mason (2000) argues that the type of family system in a society influences the onset of the fertility transition and that changes in the dominant family system can precipitate fertility decline. Systems 'that emphasize the lineage over the household or conjugal unit, found historically, predominantly in sub-Saharan Africa, and that are hierarchically structured, appear to have a higher maximum acceptable family size' (ibid.: 162) than other systems. In much of sub-Saharan Africa, labour is the main economic resource so that large families strengthen the lineage and increase the power of its leaders. The burden of caring for large numbers of descendants could also be shared across the lineage rather than falling on the shoulders of the individual household. However, the practice of polygamy in which a man has several wives, combined with the common system of the separation of male and female household budgets, leads to most of the costs of child-rearing falling on mothers rather than fathers. At the same time, since the women need children to help them work the land or otherwise assist in supporting the maternal unit, women also want large families.

Postnatal controls, such as fostering or adoption as practised by some African and West Indian families, or early marriage as previously common in China, or sending rural children to work as unpaid servants in urban households as in Peru, or as labourers as in some West African countries allow demographic smoothing across households. It may also be mutually beneficial by allowing childless women to gain status and assistance through adopting children that relatives cannot support or by providing labour in households where this is needed. The employer is expected to support the child, thus reducing the costs for the natal family, and the child may learn some skills, such as those of household management for girls or knowledge of farming or a trade for boys. In some cases rural children may also get the opportunity for education when sent to the city. However, neither parents nor children have any control over the situation and it often ends up as one of exploitation of children sexually and in terms of overwork. With the spread of HIV/AIDS and the belief that young children are safer sexual partners than adult prostitutes, sexual exploitation is increasing.

Demographic change both affects and is affected by the situation of women. On the one hand, the extent of son preference affects women's status but on the other hand, in Thailand, women's increased autonomy is seen as the explanation for one of the most rapid declines in fertility in the twentieth century (Knodel *et al.* 1987). It is also involved in the child quality versus quantity trade-off, where having fewer better-educated and healthier children is seen as preferable to many children whom the parents cannot afford to educate. Presser (2000) suggests that reliable and available contraception gives women a sense of empowerment. It is believed that about 20 per cent of married women in developing countries (100 million women) have an unmet need for family planning, but this need is most acute among illiterate couples who are often ignorant of how their bodies function (Yinger 1998).

As women increasingly move into the labour market, the stress of the double burden increases and women's leisure time becomes increasingly valued. Women in Eastern Europe and Russia were expected to be in full-time paid work while their children were looked after in state-run nurseries. When state childcare and maternity leaves were reduced after 1989, women made new choices, ranging from retreating to the home to raise their children, working from home or working part-time to allow for more family time, to spending more time in advanced education in order to improve their labour market opportunities, marrying later and choosing to have fewer children. In most of these countries total fertility rates fell from slightly more than two births per woman in 1980 to below replacement level, around 1.2, in 1997 (UNICEF 1999).

The influence of the family system and of increased education for women are now widely seen as the key to fertility decline. The Program of Action of the United Nations International Conference on Population and Development, held in Cairo 1994, concluded that:

> [I]mproving the status of women . . . is essential for the long-term success of population programmes. Experience shows that population and development programmes are most effective when steps have simultaneously been taken to improve the status of women.
>
> (United Nations 1995a: para 4.1)

The state and the body

Currently two-thirds of developing countries have national population policies or programmes (Tsui 2000). India was the first country to do so in 1951, followed by Pakistan, China, the Republic of Korea, Barbados and Fiji a few years later. However, state concerns in relation to population growth vary. In 1976, 12 per cent of 156 states thought their national fertility rate was too low, increasing slightly to 14 per cent out of a total of 170 countries in 1986 and falling to 13 per cent of 179 countries in 1996 (ibid.: 186). At the same time the number of countries with a view that national fertility was too high increased steadily from 35 per cent of countries in 1976 to 47 per cent in 1996. In 1976 just over half (52 per cent) of countries reporting had no official policy to influence fertility, while by 1996 this had fallen to roughly a third (34 per cent) and 79 per cent actively supported access to contraception (Tsui 2000). India is unusual in that in its early attempts to control the birth rate it targeted men, rewarding those men who agreed to be sterilized and enforcing this through local quotas. In most cases, however, fertility control focuses on women and coerces them to different degrees to utilize contraception and abortion. Thus state intervention in women's bodily functions has been increasing.

Three case studies show how different countries have intervened to control women's fertility and hasten or reverse the demographic transition. National attitudes to population growth in relation to economic development have varied from the 'Maoist' view that people as producers are a state's best resource to the 'Malthusian' view of people as consumers, expressed in the 1992 Earth Summit Agenda 21, which linked demographic trends to pressure on land resources leading to environmental degradation and hindering sustainable development (Quarrie 1992).

China

One of the best known examples is China's 'one-child family policy'. Between 1949 and the mid 1970s China experienced a demographic transition from high to low birth and death rates and the improved socio-economic situation for most of this period made it possible for rural families to have many children. In an early pronouncement on population Chairman Mao stated in 1949: 'Of all things in the world, people are the most precious' (Mao 1969: 1401, quoted in Milwertz 1997: 39). This remained official policy until the death of Mao Zedong in 1976. In the 1950 Marriage Law, monogamy was made the standard of sexual relations as a way of eliminating the feudal system of concubinage (Evans 1992). However, by the 1980s sexual relationships were allowed more freedom and divorce made easier, but childlessness and spinsterhood were frowned upon and the sale of child brides reappeared. Despite these changes, as Edwards (2000) makes clear, state policy on sexuality remained focused on channelling female behaviour in the service of social and moral order.

Growing imbalances between population and resources did lead to a policy encouraging planning of births as early as 1956 (Milwertz 1997). According to Milwertz (1997), the main difference in family size and structure prior to 1949 was based on class, with the rich having larger families than the poor. After 1949 the division became one based on residential location, with urban families becoming smaller, while rural families grew as child and infant mortality declined. Thus rural families were able to attain the Confucian ideal of large families with many sons. The first population policy aimed explicitly at reducing population growth was introduced in the early 1970s and encouraged later marriage, longer intervals between children and fewer children. This policy led to a very rapid decline in birth rates. Whether this decline was primarily the result of improvements in the status of women or of direct state coercion is not clear. In 1978–9 a post-Mao population policy recognized that the development of the national economy was being impeded by rapid population growth and so the one-child family policy was introduced. This policy was not applied to ethnic minorities, and many other exceptions have been allowed so that rural families with problems of labour supply for agriculture and little access to pensions are allowed two children, as are families in which the first child is a girl. The one-child policy is most strictly observed in major urban centres such as Beijing and Shanghai, where the official revaluation of girls is most widely accepted, discouraging rejection of female offspring. Coercion is most effective among workers in state enterprises, where women can be forced to abort second pregnancies by threat of loss of jobs and benefits and by imposition of heavy fines, and where their use of contraception is carefully monitored and aided by workplace birth cadres. Private entrepreneurs and richer peasant farmers are not so easily coerced and they may choose to pay fines in order to have more than one child, especially if the first child is a girl.

Women carry the main burden of this national policy of fertility control as the targets of contraception, and girls and girl babies are its victims. Over 84 per cent of couples in China use contraception, the highest level in the world after Bulgaria (UNDP 2002), with the main methods being female sterilization and the use of IUDs, both of which empower medical practitioners over women as they require professional medical attention (Edwards 2000). Sterilization is the most common method in rural areas, while it is used less frequently in urban areas. Nationally 40 per cent of couples relied on female sterilization but only 11 per cent on male sterilization. At the same time widespread improvement in medical facilities has meant that almost all births, even in rural areas, occur with medical assistance and this has led to an impressive reduction in maternal mortality to 55 deaths per 100,000 live births compared to India's 540 (UNDP 2002).

The demographic consequences of the one-child family policy are particularly noticeable in terms of the sex ratio. The imbalance caused by the national policy is reinforced by a patriarchal preference for sons. Since the 1960s the sex ratio at birth has changed from 106 boys for every 100 girls to 112 boys to every 100 girls in 1990. Sex ratios are highest, at over 120 boys to every 100 girls, in the least developed rural provinces and lowest (between 102 and 106) in provinces such as Yunnan and Tibet, where there are high proportions of ethnic minorities (Milwertz 1997).

The imbalance between boys and girls is even more marked in the case of second children (Figure 3.2). There is also under-reporting of female births, unofficial adoption of unwanted female children by childless women, female foeticide or selective abortion, especially in cities where the technology is easily available, and female infanticide (Box 2.1). In China the infant mortality rate for girls was 48 per 1,000 live births versus 35 for boys in 2000 (United Nations 2000). The shortage of women, who in 1999 made up only 48.4 per cent of the population, is now impacting their availability for marriage, leading to increased trafficking in women and their sale as brides. This is especially true for rural women, who are lured to urban areas by the promise of jobs only to find themselves sold as brides, and for disabled women, who may be sold as 'housekeepers'. Such treatment of women and girls is increasingly being seen in terms of human rights violations (Finnane 2000) and the state is attempting to stop it, even going so far as to enforce laws against it at the village level.

By 1992 China had a fertility rate below replacement and so reforms in reproductive policy were instituted, culminating in 2000 and 2001 in two major national documents institutionalizing these changes (Winkler 2002). Both documents shift the policy from simply restricting births to a broader reproductive health focus incorporating recent ideas from the Cairo and Beijing international conferences of informed choice and women's empowerment. Efforts are being made

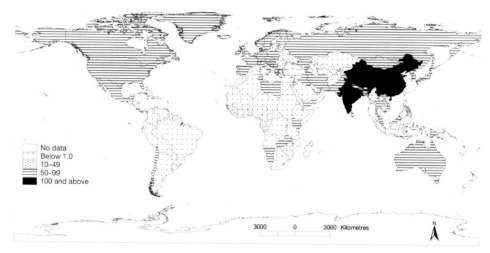

Figure 3.2 *Difference between the infant mortality rates of girls and boys (per 1,000 live births), 1995–2000.*

Source: United Nations (2000: 79–83)

to undo the unplanned occurrence of imbalanced sex ratios at birth. Emphasis is being put on incentives, such as money, for postponing marriage and childbearing and on related programmes aimed at combating poverty and providing social insurance and retirement pensions. It was estimated that, between 1971 and 1998, some 46 per cent of the decline in fertility was accounted for by socio-economic change, while the state programme had accounted for 54 per cent of the decline and births had been reduced by 338 million (ibid.). Without this decline surplus rural labour would have been one-third higher and foodgrain production would have been below self-sufficiency by 1998 (ibid.). Despite the changes, stringent birth control remains obligatory and non-compliance is still costly, requiring the payment of a 'social compensation fee'. The new rulings also ignore the creation of a 'black population' of unauthorized 'out-of-plan' children who are not entitled to government benefits. However, they do address corruption at the local level and attempt to improve the quality of the population through sex education, better health care and an innovative attention to unmarried young people and to men's health, especially the problem of erectile dysfunction, said to affect half of Chinese men over 40 (ibid.). Overall, China's birth control policies have shown a shift from the individual's duty to control population to more emphasis on her rights to quality health care.

Lack of public health education and facilities continue to threaten Romanian women's health so that, for example, they have the highest incidence of uterine cancer in Europe (Roman 2001: 57).

Singapore

Unlike the communist-controlled nations of China and Romania, capitalist Singapore used the carrot rather than the stick to influence women's fertility. This method was no more successful in influencing women's fertility decisions at the personal level (Teo and Yeoh 1999). Singapore, an island city state off the southern tip of Malaysia, has experienced remarkable demographic changes over the last three decades. During this period its population policies were designed specifically to fit development targets. The total population has increased from 2.4 million in 1980 to 4 million in 1999, while total fertility rates have fallen from 2.8 children per woman in 1970 to 1.48 in 1999. Meanwhile the dependency ratio (children and the elderly as a ratio of those of working age) was less than that of both the USA and the UK in 1999. Epidemic diseases like malaria were eradicated and by 1991 the infant mortality rate was down to 5.5 per 1,000 live births. The anti-natal 'stop at two' policy orchestrated by the Singapore Family Planning and Population Board, set up in 1966, was government policy from the early 1970s until 1987. Higher income tax relief was given for the first two children in a family, with no relief for the fourth and subsequent children. Legal sterilization and abortion were introduced in 1969, maternity leave was restricted to only two children and priority in school placements was lost after two children (ibid.). The government also manipulated housing policies in favour of smaller families, a very effective carrot in a country where 85 per cent of the population now live in modern housing provided by the state Housing Development Board (Graham 1995). The total fertility rate fell to a low of 1.42 in 1986, since when it has fluctuated but remained below replacement level.

The decline in fertility of 70 per cent over 20 years was so rapid that, in 1983, the then Prime Minister, Lee Kuan Yew, redefined Singapore's population policy as a tool in the rejuvenation and restructuring of the economy. From 1983 to 1987 Singapore's population policy entered what has been called a 'eugenics' phase aimed at selectively increasing fertility. The Prime Minister publicly supported the controversial argument that a child's intelligence is inherited and is related to the mother's intelligence. This belief led him to bemoan the low fertility of better-educated women and the growing number of unmarried female graduates (Box 3.1).

In order to rebalance birth rates in favour of the better-educated classes, the government introduced changes in primary school registration, enhanced tax relief for mothers and restructured delivery fees, all of which benefited educated mothers while at the same time providing a cash incentive for uneducated mothers with one or two children to be sterilized (Teo and Yeoh 1999). The eugenics programme provoked widespread protests and the reaction of graduate women themselves was strongly negative (Graham 1995). This laid the groundwork for the introduction of the New Population Policy (NPP) in 1987.

The NPP was a pro-natalist policy linked to the second industrial revolution, which required skilled workers for the new high-technology industries that Singapore was beginning to look to for further development. The new slogan, 'Have Three, or More if You Can Afford It', replaced the old 'Stop at Two'. Eligibility for the new package of policy initiatives was still based on examination success. The higher level of tax relief was extended to the third child and eligibility was reduced from five passes at 'O' level to three passes for women who were taxed separately from their husbands. Sterilization incentives were limited to those with no passes at 'O' level in the GCE. Tax incentives were structured to encourage the most educated to have children earlier and to reduce birth spacing. The maximum period of unpaid maternity leave for mothers employed in the civil service was increased to four years. On a visit to Singapore in 1988 the immediate positive reaction to these incentives was visible. Obviously pregnant women were very noticeable on the streets and the local papers contained complaints from senior civil servants of the inconveniences caused by the sudden departure of unexpectedly large numbers of women on maternity leave.

One aspect of this selective family planning not officially mentioned in the policy initiatives was the differential effect on Singapore's three major ethnic groups: Chinese, Malays and Indians. The Malay and Indian populations have, since the late 1970s, had higher total fertility rates than the Chinese population, but a study in 1992 appeared to indicate that all three ethnic groups had increased their fertility rate since 1987 but that only the rate for Malays was above replacement level. The general growth of education and employment

Box 3.1

The impact of education on fertility in Singapore, Sri Lanka and the Middle East

The Social Development Unit (SDU) of Singapore is a state-run marriage bureau initially set up in 1985 to find partners for women graduates. The need for such an organization first became apparent in 1982, when ministers studied the returns of the 1980 census. They found that a large proportion of female graduates were not married. It was realized that the traditional Chinese males' habit of preferring wives less educated than themselves was creating difficulties for a new generation of women who had graduated from university. In 1984 the National University of Singapore adjusted its entrance requirements to make it easier for men to get in. Men had found it more difficult than women to meet the second language requirement and the sex ratio of students had begun to swing in favour of women. Since 1979 the university has limited the number of women admitted to the prestigious medical faculty to one-third of each year's intake. While Singapore men want to marry women with less education than themselves, Singapore women will not accept men of inferior education and so 39 per cent of female graduates remain single. The result is not only that a large number of female graduates are unmarried but also that 38 per cent of men without higher education fail to get married. For a Singapore government sensitive to fertility rates among a small population, the further worrying conclusion was that some of the island's brightest women were failing to reproduce.

The task of the SDU is to bring together single male and female graduates in order to enhance their opportunities of finding a mate. In its first three years of operation the SDU succeeded in finding marriage partners for 400 graduates. The agency fills a gap created by the pressures of life in modern Singapore. Jobs are markedly gender-segregated so that, for example, teachers tend to be women and engineers tend to be men, with very little opportunity for meeting each other. In addition, working men and women return home so tired after what often amounts to a 12-hour day, that they have no time to plan social activities in the evening. By the start of the third millennium, the SDU had extended its mandate to include non-graduates who could take advantage of a separate programme of activities run by another government agency, the Social Development Service (SDS), through which they could meet marriage partners, often through computer dating.

The subservient role of women, most noticeable in Japan, Korea and Taiwan, has its roots in Confucian culture. In Korea it is a tradition that, when a girl marries and is about to live with her husband, her mother gives her a stone saying 'even if your husband and mother-in-law provoke you, you only open your mouth when the stone starts to speak'. A measure of women's new independence came in replies given to a recent survey of public opinion in Singapore. Ninety per cent of men said they thought marriage was necessary for a 'full life'. Only 80 per cent of women believed this to be the case.

In Sri Lanka marriages are still arranged by parents but the pool of suitable partners is being extended through the use of newspapers. Advertisements for marriage partners in the main English-language newspaper, the *Sunday Observer*, available on the paper's website, show a growing interest in matching educational levels as well as looks, religion, caste and class:

> A qualified partner is sought by Buddhist Vishwakula parents for 29 plus 5'2" slim attractive Lawyer daughter with assets and excellent character attached to a firm. Caste immaterial. Good horoscope.

> Singhalese parents seek suitable partner for their daughter lives in Canada, 31 years old and 5'7" tall, working professionally in Canada she owns her own modern house and car. Religion and race immaterial, but should be qualified and willing to live in Canada.

> Singhalese parents seek Born-Again Christian, well educated partner, below 32 years, willing reside in Australia for graduate 25 year daughter.

> Professionally qualified partner holding responsible position sought by Colombo suburbs respectable Govi Buddhist parents for their only child daughter 27, 5'3", Attorney-at-Law and Bachelors Degree holder, both passed with Honours in English medium. Attended leading Colombo school very fluent in English, draws high salary and inherits modern furnished house new car and other valuable assets. Horoscope.

In the United Arab Emirates the government does not want its citizens to marry foreigners. The government-financed UAE Marriage Fund's policy is that mixed marriages, especially if they involve a non-Muslim, threaten social stability. Some 28 per cent of the country's one million people are married to a foreigner and 79 per cent of UAE men who divorce local wives go on to marry a foreign one. This has caused an excess of local spinsters so the fund recently offered a special financial premium for citizens who marry 'older', that is over 30 years of age, local women. The Marriage Fund was set up a decade ago to assist men who could not afford to marry. In addition to direct financial assistance the Fund further reduces the costs of weddings by organizing mass weddings for dozens of couples throughout the country.

The pattern of divorce reveals a contradiction at the heart of UAE society. Only 4 per cent of divorced local women have finished secondary school and only 1 per cent have a degree. The UAE encourages education for women but custom demands that women marry young and often to a cousin. Although the country wants to have the benefit of educated women, women are expected to marry before they have had time to complete their education.

In all three countries, Singapore, Sri Lanka and the United Arab Emirates, educated women find it hard to meet suitable husbands. This has become a national problem, with governments setting up agencies to promote such marriages in Singapore and the UAE and parents and state agencies utilizing electronic methods to extend the pool of acceptable partners.

Sources: adapted from: *Economist*; *Financial Times*; *Sunday* Observer (2002); SDU and SDS websites, www.sdu.gov.sg and www.sds.gov.sg (accessed 4 October 2002).

for women, fostered by the expansion of service and hi-tech industries, had increased the opportunity costs of having children. The increase in individualism as illustrated by the decline in the use of traditional matchmakers and the new freedoms in the use of marriage partners, including computer dating (Box 3.1), are making fertility an increasingly private decision (Graham 1995). Teo and Yeoh (1999) argue that the pro-natalist NPP has met with far more critical responses compared to the pliancy characteristic of the response to the old anti-natalist policy. According to Phua and Yeoh (2002) it is now clearly accepted that the procreation problem remains acute.

The alliance between the Vatican and many Muslim states at the United Nations Cairo Population Conference in 1994 and the Beijing Conference in 1995, against empowering women in terms of reproductive rights, underlines the continuing interest of governments in controlling women's bodies. Even in 2002 the government of Peru, despite having a President who had only just recognized a teenage illegitimate daughter after considerable public pressure, refused to accept aid from both Spain and Britain for improvements in women's reproductive health care services, under pressure from ultra-conservative Roman Catholic groups (AWID 2002c).

Education

As Box 3.1 shows, the educational levels of men and women affect many life options. In 1960 only 58 per cent as many women as men were literate in the developing world. By 2000 female adult literacy was 81 per cent of male literacy and youth literacy was 91 per cent, reflecting recent increases in access to education for all in developing countries (UNDP 2002). However, in South Asia the figures were 66 per cent of male adult literacy and 79 per cent of male youth literacy, while in Latin America and the Caribbean the corresponding figures were 98 and 101 per cent in 2000 (ibid.). In several countries in the Middle East (United Arab Emirates and Qatar), in two Caribbean nations (Bahamas and Jamaica), in Nicaragua and in Botswana and Lesotho, female adult literacy was higher than male adult literacy (ibid.). Lesotho in southern Africa had the highest relative female adult literacy in the world at the beginning of the new millennium, at 129 (ibid.), as explained in Box 2.2. In general, poor parents actively seek education for their children as the best means of improving their income-earning options, but overburdened mothers may be forced to take daughters out of school to assist with childcare and household chores.

Figure 3.3 shows the international pattern of relative male and female illiteracy rates in 1999. In the industrialized world of the North there were virtually no gender differences in literacy and this was true also of South Africa and the Philippines and most of the countries of Latin America and the Caribbean. Women had the lowest literacy rates as compared to men in West Africa and South Asia but more women than men were enrolled in tertiary education in most countries of Central and Eastern Europe, Latin America and the Caribbean, and the Middle East, with the greatest recorded imbalance being in Barbados in 1998 where the female-to-male ratio was 228 (ibid.). However, UNESCO estimated that there were 113 million children not attending school in 1998 of which 60 per cent were girls (World Bank 2001). Progress in increasing enrolment rates has been least satisfactory in sub-Saharan Africa, where the number of school-age children is expected to increase the most, by 33 million between 2000 and 2015, and teachers are being lost to HIV/AIDS (ibid.). Women achieve lower levels of education than men in the majority of developing countries because of distance between home and school and lack of transport, which may make it dangerous for girls to travel to school. The costs of school in terms of the loss of the child's labour at home and the financial burden of paying for school supplies, suitable clothing such as uniforms, school fees and bribes to teachers also means that parents may decide not to educate daughters. This is particularly so where marriage systems place the bride in the hands of the husband's family so that her natal family may feel that educating a daughter is investing in someone else's family. Girls may also be vulnerable to sexual harassment in schools and so may jeopardize their marriage potential. In rural schools facilities are often poor and there are few women teachers, making such schools especially unattractive to girls. For older women it may be hard for them to attend adult literacy classes because of the demands of their reproductive tasks. Without literacy women may not be aware of their legal rights and may be unable to benefit from opportunities for further training.

However, the gender balance in higher education in the developing world has improved very rapidly. Between 1965 and 1985 estimated female enrolment in tertiary education in southern Yemen and Qatar increased from 0 to 19 and 57 per cent of the total respectively; in

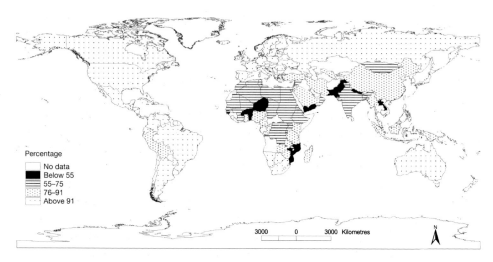

Figure 3.3 *Adult literacy: female rate as a percentage of male rate, 1999.*
Source: United Nations (2001: 218–21)

Guatemala it increased from 9 to 28 per cent and in Brazil from 25 to 48 per cent. In Latin America as a whole more women than men were enrolled in tertiary education by the end of the 1990s, except in Costa Rica, Honduras, Mexico and Peru. However, access to higher education is strongly dependent on class, location and income. Within the education system women tend to be channelled into certain subject ghettos, such as nursing, education and social work, while the courses leading to the best-paid jobs, such as medicine, law and engineering, are still dominated by men. The expansion of women in tertiary education, as has occurred in the industrialized North, is related to the expansion of jobs for educated women and, especially in the Middle East, to a growing interest by men in better-educated wives (Box 3.1).

Social reproduction

Activities carried out to maintain and care for family members are generally ignored in national accounts, but they are essential economic functions which ensure the development and preservation of human capital for the household and for the nation. Education of young children and organization of the household so that members can maximize their access to educational facilities may be part of these activities. Others may include fuel and water collection, care of children, the sick and the elderly, washing clothes and processing, preparing and cooking food. For most families throughout the world these jobs are done by women.

Much of the research focusing on women in the South has looked at biological reproduction in isolation from commodity production and has ignored social reproduction. Women perform the great bulk of domestic tasks in all societies. Even in Cuba where, by statute, men are supposed to assist women in such work, 82 per cent of women in the capital city, Havana, and 96 per cent of the women in the countryside have sole responsibility for domestic chores. The equivalent figure in Britain in 1988 was 72 per cent. In subsistence societies, the separation between reproductive and productive tasks is to a large extent artificial, symbolized by the woman with her baby on her back working in the fields (see Plate 3.1 and Table 3.1).

Plate 3.1 *China: women preparing tobacco leaves for curing in a village in Yunnan, in the south-west. Note the small child being carried on the back of one of the workers. The actual drying and curing of the tobacco is done by men.*

Source: author

Table 3.1 *Sri Lanka: gender roles in household activities*

Activity	Percentage of time required for each activity by gender	
	Male	*Female*
Food preparation	8	92
Winnowing and parboiling rice	0	100
Preserving food for the hungry season	20	80
Storing grain at harvest time	70	30
Growing of fruits, tubers, greens and vegetables for home consumption	20	80
Caring for animals (goats, cows, buffaloes)	50	50
Milking cows	100	0
Fetching water	2	98
Collecting firewood	35	65
Care of house and yard	5	95
Childcare responsibilities	10	90
Bathing children	20	80
Care of sick and elderly family members	10	90
Participating in village ceremonies	55	45
Participating in village social activities	90	10
Participating in community development activities	95	5

Source: Anoja Wickramasinghe (1993: 170; adapted from Table 13.5).

Gender and time use

Time use data is problematic in that it is only available for a few countries, is rarely disaggregated by both gender and urban/rural residence and methods of measurement vary from survey to survey. However, it appears that in almost all countries women work more hours than men. As women increase their paid workhours they find that in most cases men do not increase their share of the unpaid burden of childcare and housework. There is some indication that rural women put in longer hours than their sisters in urban areas because of the poorer level of household services in the countryside, such as running water, electricity and gas for cooking (Table 3.1). UNDP figures for Kenya indicate that urban women put in 103 per cent of men's worktime and 135 per cent in the rural areas (590 minutes per day for women and 572 for men in urban areas versus 676 in rural areas for women and 500 for men) (UNDP 2002). In Nepal women work 105 per cent of men's work time in the cities

Plate 3.2
Ghana: a woman collecting firewood in a cassava field.

Source: author

Plate 3.3
Ghana: women processing cassava for gari.

Source: author

Plate 3.4
East Timor: young women carrying water back to their village.

Source: author

and 122 per cent in rural mountainous areas (579 versus 554 minutes per day in urban areas and 649 versus 534 minutes in rural areas) (ibid.).

In sub-Saharan Africa women spend an estimated four hours a day on collecting firewood and water, childcare and preparing food (see Plates 3.2 and 3.3). This is in addition to the time they spend on agriculture, craftwork or trading. At certain times of year the domestic tasks will take longer than normal. In the dry season the village well may run dry and so women have to walk further for daily supplies of water (see Figure 3.4 and Plate 3.4). If the nearby source of firewood is over-utilized, women will have to spend longer searching for fuel and perhaps also reduce the amount of cooked food that they prepare. At planting and harvesting periods, when more time must be spent in the fields, domestic household tasks must be reduced.

Figure 3.4
Women's use of time.

Source: The International Women's Tribune Centre

– BUT WHAT WILL WOMEN DO
IF THEY DON'T HAVE TO
CARRY WATER FOUR HOURS
A DAY?

Families with several small children absorb much of women's time in childcare unless there are older siblings who can assist the mother, although she may not wish them to do so if it means giving up their opportunity to attend school. The nutritional level of children is often negatively related to the distance mothers have to walk to collect water. The average round trip from house to water supply in Africa is five kilometres and thus the effort of carrying water can absorb 25 per cent of a woman's calorific intake. Domestic chores in developing countries, where household appliances are rare, consume a high proportion of women's energy and time (Table 3.1). The recent introduction of village grain mills in The Gambia was found to save women 60 to 90 minutes per day. But it was the saving in energy that was most appreciated, as the pounding of sorghum is hard work.

Housing

Housing conditions have an effect on the time and effort consumed in housework and on the health and well-being of residents. Between 1990 and 2000 access to improved sanitation facilities in urban areas improved from 68 to 79 per cent for the urban population in low-income countries and from 75 to 82 per cent in middle-income countries (World Bank 2001). In rural areas the situation was much worse. A wooden shack with an earthen floor suffers from dust in the dry season and mud in the wet season (Chant 1984). Cleaning is never-ending and it is difficult to see any positive results. Residential areas without electricity, piped water, paved roads or sewers make both housework and childcare harder and more time-consuming (Chant 1987). Lack of services is a still a major problem in rural areas. In Brazil in 1991, 88 per cent of urban homes had piped water but only 9 per cent of rural homes, while 45 per cent of urban dwellings had sewer connections compared to 2 per cent of rural homes (IBGE 1994). Similar differences are seen in the use of fuel with 83 per cent of urban homes but only 14 per cent of rural using gas, while 77 per cent of rural homes used wood for cooking versus 11 per cent of urban dwellings in Brazil in 1980 (ibid.). Today residents of most cities in the Central and Eastern European countries, in South-east Asia, North Africa and the Middle East have access to potable water and to sewage, electricity and telephone connections. In some parts of sub-Saharan Africa, as in Niamey (Niger), Kinshasa (Congo Democratic Republic) and Brazzaville (Republic of Congo), and in Apia (Samoa) there are no sewage lines. Even in Asunción, capital of Paraguay, only 8 per cent of homes have sewage connections, and in Dhaka, capital of Bangladesh, only 22 per cent have a public sewage system, while most other cities in the country have none (World Bank 2001). Where fixed telephone lines are in short supply, as in Dhaka, Asunción, Tripoli (Libya), much of sub-Saharan Africa and several Latin American cities, cellphones are an increasingly popular alternative, as they are in rural areas of most developing countries.

The household

The size and make-up of the household determines to a large degree the burden of work on women. It has been shown that, in nuclear families, the full burden of social reproduction falls on the wife and mother, but in extended and female-headed households there is much more sharing of tasks because the mother has more autonomy. In the developing world the nature of households is changing very rapidly. Co-residential households may not necessarily be child-rearing units, nor are they always economic units and the role of non-residential migrant members may be crucial to an understanding of the function of the household.

It is usually assumed that the household head is male and that he allocates household labour and organizes the distribution of consumption goods among household members so that all benefit and participate equally. Clearly this is not always so and much depends on gender relations and power within the household. In general, women's duties are closely associated with the collective aspects of family consumption, while men have more individual control over their own personal consumption of resources. In order to understand the role of the household in development, it is essential to recognize the dynamics of the system: the changing nature of production, distribution and consumption relations within the household, and the effect of life cycle on dependency ratios.

Learning outcomes

- Most of the work undertaken in the social and biological reproduction of the family is done by women but controlled by others.
- State population policies may determine individual reproductive behaviour.
- The influence of women's education level on birth rates and marriage.
- The time demands of the triple reproductive roles of women.

Discussion questions

1 Why does fertility decline with urbanization and education of women?

2 Rehearse the ways in which housing and family structure can influence women's use of time in domestic tasks.

3 Elaborate on the reproductive roles of women.

4 How can state policies influence fertility rates?

Further reading

Croll, Elizabeth (2000) *Endangered Daughters: Discrimination and Development in Asia*, London: Routledge. Provides an overview of discrimination against young women, especially in India and China.

Graham, Elspeth (1995) 'Singapore in the 1990s: can population policies reverse the demographic transition?', *Applied Geography* 15 (3): 219–32. An assessment of the history of Singapore's demographic planning

Kligman, Gail (1998) *The Politics of Duplicity: Controlling Reproduction in Ceauşescu's Romania*, Berkeley and Los Angeles: University of California Press. A detailed analysis of Ceauşescu's population policy and its effect on women and children.

Websites

www.cedpa.org   CEDPA (The Center for Development and Population Activities) focuses on issues of reproductive health and on women's empowerment at all levels of society to be full partners in development.

www.hdr.undp.org   United Nations Development Programme's *Human Development Report 2002.*

4 Gender, health and violence

Learning objectives

When you have finished reading this chapter, you should understand that:

- health problems vary with life stage
- gender patterns of HIV/AIDS infection are changing
- violence is a health problem with social, economic and political causes
- changing men's attitudes is very important.

Spatial differences in gendered health problems have rarely been considered but are becoming increasingly complex as international flows of population increase (Dyck *et al.* 2001). An epidemiological transition is now under way in all regions of the world, indicating a shift from a predominance of infectious and parasitic diseases to one of chronic and degenerative diseases. But this is not a linear unidirectional change and counter-transitions also occur (Salomon and Murray 2002). Many developing countries, and countries in transition, are confronting a double burden of fighting emerging and re-emerging communicable diseases, such as HIV/AIDS, tuberculosis and malaria, in parallel with the growing threat of non-communicable diseases, such as those caused by increasing use of alcohol and tobacco. In 2002, according to the World Health Organization, the three leading health problems, measured in disability adjusted life years (DALYs), in low-income countries were malnutrition (14.9 per cent), unsafe sex (10.2 per cent) and unsafe water, sanitation and hygiene (5.5 per cent), while in middle-income countries the most common causes of poor health were alcohol (6.2 per cent), high blood pressure (5.0 per cent) and tobacco (4.0 per cent) (Dyer 2002). Indoor exposure to smoke from solid-fuel fires was the fourth cause of poor health in low-income countries, especially for women, while in middle-income developing countries malnourishment and obesity were the fourth and fifth leading health problems. Currently 30,000 people die each day in developing countries from infectious diseases but world trade rules on drug patents restrict poor people's access to essential medicines (Pearson 2002). Such access to medicines is gendered, as women's experience of illness differs from that of men and women generally have little decision-making role at household or national level.

Poverty and health are closely related, but economic improvement does not necessarily lead to better public health. Increased cost of health care, as occurs under structural adjustment policies, has immediate effects on attendance at clinics, but may stimulate interest in family planning as a way of cutting costs by reducing the number of children born (Iyun and Oke 1993). For poor women the suffering is especially marked because of their low social status, few decision-making rights, their heavy workload, including family health care, and their experience as bearers of children. For men the loss of status through unemployment may lead to the adoption of life-threatening behaviour, such as alcoholism and drug-taking and even suicide.

In the transition countries collapse of state health services, and the need to pay for medical care and to be proactive in seeking care have led to generally declining health and the emergence of resistant strains of diseases such as tuberculosis. The increase in smoking, alcoholism and related violence and crime, has led to an actual decline in men's life expectancy. This decline was less than one year between 1990 and 1997 in Bulgaria, Romania, the Baltic countries, and Turkmenistan, less than two years in Kyrgyzstan and Tajikistan (1990 to 1995), but three to four years in Russia, the Ukraine, Moldova (1990–5) and Belarus. However, male life expectancy rose in the richer Eastern European countries soon to join the European Union: Poland, the Czech Republic, Slovakia, Hungary and Slovenia, and also in Armenia and Azerbaijan (UNICEF 1999: Figures 2.2 and 2.3). Female life expectancy at birth was less affected, falling by less than two years in Russia, Ukraine, Belarus, Moldova and in most of the Central Asian countries (ibid.: Figures 2.2 and 2.3). The effect of improved maternity care and declining fertility rates is shown by declines in female mortality rates in most countries in transition for women aged 20–39, except in the eight countries of Latvia, Lithuania, Belarus, Moldova, Russia, Ukraine, Azerbaijan and Kyrgyzstan, where the increases were less than one per cent (ibid.: Figure 4.1). Among men of the same age group male mortality rates increased in 11 countries, with increases of over 1 per cent in Belarus, Russia, Ukraine and Kazakhstan (ibid.: Figures 2.2 and 2.3). By the beginning of the twenty-first century, women could expect to live more than 10 years longer than men in the Baltic States, and in the Ukraine, Belarus and Kazakhstan, reaching an extreme of 13 years longer in Russia (Figure 2.3), where male life expectancy has fallen from 64 to 59 years since 1970 and women's from 74 to 72 years (Figure 2.2).

Despite the general improvement in women's life expectancy when compared to that of men in the last three decades, women experience more chronic, debilitating diseases than men, while men, on the whole, live shorter but healthier lives than women. Life expectancy is shortest in sub-Saharan Africa, where maternal mortality and AIDS are more prevalent. In Botswana, where almost 40 per cent of the adult population is infected with AIDS, general life expectancy has recently fallen below 40 years for the first time since 1950 (McGregor 2002: 3). Declines in life expectancy due to HIV/AIDS are also seen in Kenya, Malawi, South Africa, Swaziland, Uganda, Zambia and Zimbabwe. HIV/AIDS is now the leading cause of premature death and loss of health from disability for women globally (Krug et al. 2002).

Losses due to civil wars are reflected in the reduced life expectancy figures for both men and women in the Balkans, Rwanda and Somalia as civilian deaths increase in modern warfare. Women and children make up the majority of refugees from war and famine. Health care in refugee camps is often inadequate and women may have to trade sexual favours for food and protection from violence (Harris and Smyth 2001). It is estimated that among the 'boat people' fleeing Vietnam in the late 1970s and early 1980s, 39 per cent of the women were abducted and/or raped by pirates while at sea (Krug et al. 2002: 156).

Although not officially recognized as health workers, women are responsible for 70 to 80 per cent of all the health care provided in developing countries (Pearson 1987; Tinker et al. 1994). Therefore, improving their own health and educating them to detect and prevent infectious diseases and to practise proper hygiene and nutrition is a cost-effective way to improve family health. This role is so important that it affects children's survival and education levels. When a mother dies, the mortality of small children rises and older children spend less time in school. Such effects are not significant when an adult male dies.

Nutrition

Poor rural households may spend as much as 90 per cent of their income on food and yet may still eat an inadequate diet. Women's lower status relative to men and their biological role in reproduction often puts them at higher risk than men for many nutritional problems. Women work longer hours than men virtually everywhere. In many countries of the South they are last to eat in the family and so may have to survive on less food in terms of both quality and quantity (Pryor 1987). Thus they are often overworked and underfed. Poor nutrition makes people more susceptible to disease.

Seasonal fluctuations in food supply also affect rural nutrition. At the time for planting crops there is an increased demand for labour but food supplies from the previous season are depleted and, because of rainfall, mosquito-borne diseases are more prevalent. During this hungry season the nutritional status of family members is at its lowest and meals are often cut out. In north-eastern Ghana it was found that for most of the year 30 per cent of men and 50 per cent of women were underweight but in the hungry season these figures increase to 49 per cent and 63 per cent respectively (Awumbila 1994).

In Bangladesh, surveys in 1996–7 revealed that 52 per cent of women (36 per cent in urban areas and 54 per cent in rural areas) were affected by chronic energy deficiency due to inadequate food (Marcoux 2002). But Marcoux concludes that, although many girls and women are undernourished, there is no clear picture of consistent gender discrimination in nutrition in recent national surveys (ibid.). Between 20 and 45 per cent of women of childbearing age in the developing world do not eat the World Health Organization's recommended amount of 2,250 calories per day under normal circumstances, let alone the extra 285 calories a day needed when they are pregnant (ICRW 1996). Maternal infection is worsened by deficiencies in iron and Vitamin A. Lack of iodine in the mother's diet can result in mental retardation in the child. Iron-deficiency anaemia in women increases the risk of miscarriage and having low-birthweight babies and makes women more susceptible to diseases such as malaria, tuberculosis, diabetes, hepatitis and heart disease. Anaemia affects over two-thirds of pregnant women in low-income countries (79 per cent in South Asia, 54 per cent in East Asia and the Pacific, 45 per cent in sub-Saharan Africa and 34 per cent in Latin America and the Caribbean).

Health and life stage

Children under five are dependent on their mothers for health care. Mothers with secondary education are more likely to get their children vaccinated, and urban families are more likely to take their children for medical treatment because doctors and hospitals are more accessible than in rural areas. In cultures where girls have less status than boys they are treated less well. In India, girls were found to be four times more likely than boys to suffer acute malnutrition and 40 times less likely to be taken to a hospital when sick. In India and China more girls than boys die before their fifth birthday, despite girls' biological advantage (World Bank 1993). Early lack of protein-energy foods leads to permanent stunting and incomplete brain development. It has been estimated that two-thirds of the adult women in developing countries in the mid 1980s were stunted, or suffered from the effects of iodine deficiency, or were blind because of Vitamin A deficiency caused by inadequate childhood nutrition (Merchant and Kurz 1992). However, child malnutrition has declined globally. In 1970 as many as 46.5 per cent of children under five in developing countries were underweight. By 1995 this proportion had fallen to 31 per cent (167 million) (Smith and Haddad 1999) due largely to improvements in women's education, which explained 44 per cent of the total reduction. A further 26 per cent of the decline was due to improvements in food availability, and improved access to pure water explained 19 per cent. Since there was little improvement in women's status over the period 1970–95, only 12 per cent of the change could be attributed to this factor. Smith and Haddad's (1999) project, using a status quo scenario, estimated that 18 per cent of children in developing countries will be malnourished in 2020. They also assumed there would be a regional shift in the distribution of these children, with the proportion in South Asia falling from 51 to 47 per cent, but sub-Saharan Africa's share rising from 19 per cent to nearly 35 per cent. Improvements depend primarily on increasing food availability and women's education and, in South Asia particularly, raising women's status relative to men (ibid.: 1999).

Prepubescent girls may also be exposed to sexual abuse. The World Health Organization (Krug *et al.* 2002) sees both child marriage and female genital mutilation as forms of sexual violence affecting young girls. Studies in Lima, Peru and Malaysia revealed that 18 per cent of the victims of sexual assault were aged nine or younger and a study in Nigeria found that 16 per cent of the female patients seeking treatment for sexually transmitted diseases (STDs) were under the age of five (Tinker *et al.* 1994). Such situations are increasing as the male belief that intercourse with a young child protects the assailant from STDs and AIDS spreads.

Female genital mutilation (FGM), also known as female genital cutting (FGC), or female circumcision, is carried out on about two million young girls every year, mainly in 28 African and Middle Eastern countries. It is estimated that approximately 135 million females in the world have undergone genital mutilation (www.amnesty.org). FGM is justified on the basis of a belief that by reducing women's physical ability to enjoy sex they will be less likely to be unfaithful to their partner. Thus uncircumcised women are often considered unacceptable as wives. After the operation, to reduce urination, they are not allowed to drink, despite the heat, for at least 24 hours. This removal of the external genitalia, almost always without anaesthesia and with non-sterile instruments, can cause acute pain, recurrent urinary tract infection and difficulties in childbirth due to obstruction from scar tissue. Several African countries have passed laws forbidding FGC but they have not been strongly enforced and may have pushed the practice underground. The growth of public rejection of this cruel initiation rite is illustrated by the arrests of ten women in Sierra Leone in July 2002, following the death of a 14-year-old girl during a ritual ceremony involving genital cutting (*New York Times* 2002). Young women in Kenya have begun to resort to taking refuge in churches, or to lawsuits against their parents to avoid FGC (IRINnews 2003). At a conference on the topic in Ethiopia, the wife of the President of Burkina Faso called genital cutting 'the most widespread and deadly of all violence victimizing women and girls in Africa' (Lacey 2003: A3). Even without FGC, malnourished and very young mothers tend to produce underweight babies with low survival rates and to have difficult births because pelvic bone growth is not completed. Thus poverty transmits poor health from generation to generation.

Adolescence

Adolescents are generally healthy but early marriage leads to early childbearing before their bodies are mature, which causes long-term negative health effects. 'For many young women, the most common place where sexual coercion and harassment are experienced is in school' (Krug *et al.* 2002: 155). This appears to be most common in Africa, where teachers are often the instigators of such violence. However, in countries as different as Nigeria and Lesotho it has been reported that local chiefs come to schools in search of new young wives. Young girls and boys may go into prostitution as a way of surviving, but adolescents are biologically more vulnerable to STD and HIV infection than older women and men (see Chapter 8). Young men and women may also be ignorant of the basic processes of human reproduction because of lack of sex education, as Harris (2000) has shown for Tajikistan. Ignorance of human physiology, lack of communication on sexual matters between partners and the low status of adolescents make use of preventive measures very limited (Weiss and Gupta, 1998). Risks associated with abortion are much higher for girls under 16 than for older adolescent mothers.

Adulthood

While infant mortality rates have fallen by half over the last two decades there has been little change in maternal mortality rates in countries of the South. Almost 99 per cent of maternal deaths take place in developing countries and the rates of maternal mortality in rich and poor countries show a greater disparity than any other public health indicator (Figure 4.1). Each year half a million women die from direct complications of pregnancy and childbirth – one woman every minute of every day. Reasons for maternal mortality include malnutrition, the early age at which women begin childbearing, inadequate spacing between births, total number of lifetime pregnancies, and the lack of medical care for high-risk pregnancies. Shortage of adequate medical facilities and trained midwives is especially a problem in Afghanistan, where maternal mortality, at 1,700 per 100,000 births, is probably the highest in the world. In 2003 90 per cent of hospitals in Afghanistan lacked equipment to perform Caesarean sections and 70 per cent of health clinics could not provide basic maternity services (Miller 2003). Abortion is another major cause of maternal mortality in poor countries, where women do not have access to family planning. An estimated 100,000

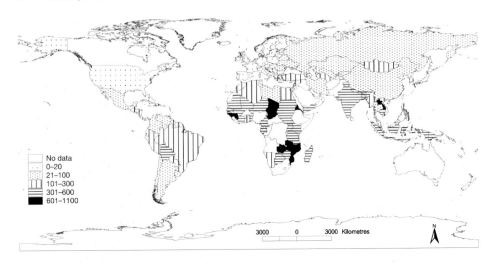

Figure 4.1 *Maternal mortality rate (per 100,000 live births) any year between 1980 and 1998.*

Source: United Nations (2000: 79–83)

to 200,000 women die each year from illegal abortions, mostly carried out under insanitary conditions. Women in sub-Saharan Africa bear three times as many children as their US counterparts and have maternal mortality rates 70 times those of the USA. However, recent research has shown that having many children and breastfeeding them for six months or more reduces the risk of breast cancer. Between 1980 and 1990 maternal deaths per 100,000 live births fell in South Asia from 731.4 to 312.2, and in Arab States from 456.6 to 297.2, but in Latin America and the Caribbean, Southeast Asia, Oceania and East Asia they increased (ICRW 1996).

Old age

Populations are ageing the world over but the majority of the almost 500 million women aged over 50 live in developing countries, and more women than men are elderly. By 2020 one in five women in developing countries will be over 50 years of age. The projected 250 per cent increase in this age group has major implications for women's health and well-being, not only for the elderly but also for their caregivers, who are predominantly also women (INSTRAW 1999). As long ago as 1982 the United Nations held a World Assembly on Ageing, and in 1987 the United Nations International Institute on Ageing, United Nations Malta, was set up and has been conducting courses ever since. Comments from developing countries' course participants repeatedly complain of lack of infrastructure and trained personnel in their countries for the rapidly growing number of older persons (ibid.).

Years of malnutrition mean that older women suffer from chronic health problems, such as diabetes, arthritis, cancer and heart problems. Loss of visual acuity and hearing make life difficult and most cannot afford spectacles and hearing aids. In many cultures the elderly are respected and can depend on care from their children but this is changing with the growth of nuclear families and migration. Since women usually marry older men they are likely to be left alone as widows and may be abandoned and become destitute. Globally, 49 per cent of deaths of men but only 27 per cent of deaths of women over 60 years of age, in low- and middle-income countries, are due to suicide (Krug et al. 2002). Regionally, suicide rates among the elderly in poor countries are highest in East Asia and the western Pacific and lowest for men in the eastern Mediterranean and for women in the Americas (ibid.).

In some cases older women may develop new roles caring for their grandchildren (see Plate 4.1) as the children's mothers migrate in search of work overseas, as in the West Indies or the Philippines, or die of AIDS, as in much of sub-Saharan Africa. This is hard work, but in the case of migrants it does help to ensure a regular income for elderly women. Where the children are left as orphans, the grandmother-headed family will find it hard to produce adequate food because of lack of labour and so the whole family will be threatened with starvation.

Many grandmothers wield great influence on maternity and child-feeding practices (Awumbila 2001). In order to improve these it is necessary to provide training for grandmothers, as well as young mothers. It is also important to give information on nutrition, as many of these grandmothers may become chief caregivers for their grandchildren. Such training recognizes the importance of their role, so empowering older women, and it has been found to have a major effect on community health practices (Aubel et al. 2001)

Gendered aspects of major diseases

Malaria and tuberculosis are becoming resistant to existing medications and becoming more devastating in poor countries.

Plate 4.1
Thailand: a grandmother carrying a child on her back in a northern village.
Source: author

Malaria causes one million deaths a year and costs Africa more than US$12 billion each year. It could be prevented through making treated bednets cheaper and more widely available, but such simple preventive measures have not been fully implemented despite many international pledges of aid in 2000. However, over the last decade HIV/AIDS has become a major health problem in most parts of the developing world (Figure 4.2). It was originally considered a disease of male homosexuals and injecting drug users, but it is now clear that heterosexual transmission is more common, accounting for 70 per cent of infections (UNAIDS 2000). HIV/AIDS infection is also unusual in that it is positively correlated with economic status, which is a reversal of the normal relation between income and health. Women are biologically more susceptible to infection and some empirical evidence shows the rate of transmission from male to

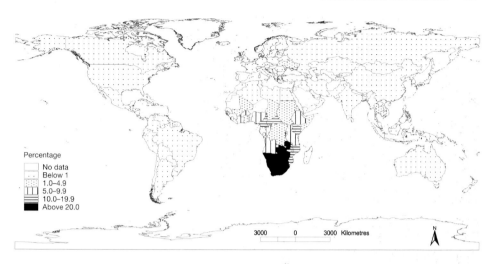

Figure 4.2   *Percentage of population with HIV/AIDS, end of 2001.*
Source: UNAIDS (2002: 190, 194, 198)

female to be two to five times higher than from female to male. Women can pass the disease on to their children in childbirth, with 3.2 million children having AIDS in 2002. Of the 4.2 million people newly infected in 2002, million are women, and women accounted for 1.2 million of the 2.5 million deaths from AIDS. At the end of 2001 it was estimated that 38.6 million adults and children were infected, of whom 19.2 million are women (UNAIDS 2002). In sub-Saharan Africa, where 71 per cent of those infected live (Figure 4.2), 58 per cent of HIV positive adults were women. There was also a female majority among infected adults of 55 per cent in North Africa and the Middle East, and 50 per cent in the Caribbean (Figure 4.3). Given the devastating effect on children's health of the loss of a mother, it is especially depressing to see this change in the gendered pattern of infection (ibid.). 'HIV/AIDS is not only driven by gender inequality – it entrenches gender inequality, putting women, men and children further at risk' (Bell 2002). The dominant risk factor is now heterosexual sex.

Botswana has the highest HIV/AIDS infection rate among adults aged 15–49 (38.8 per cent), followed by Zimbabwe (33.7 per cent), Swaziland (33.4 per cent) and Lesotho (31.0 per cent) (Figure 4.2). Of the 14 million children under 15 years of age orphaned by AIDS, 79 per cent were living in sub-Saharan Africa at the end of 2001 (UNAIDS 2002). Sex between men and boys occurs widely, especially in situations where access to women is limited, as in

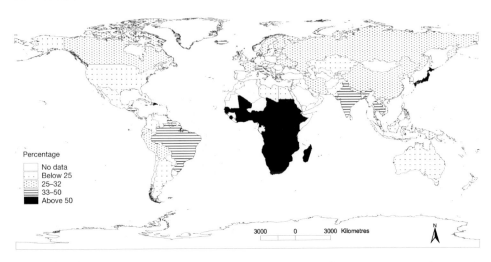

Figure 4.3 *Percentage of women (ages 15–49) in the HIV/AIDS population, end of 2001.*
Source: UNAIDS (2002: 190, 194, 198)

all-male schools and prisons. Rates of homosexual sex range from 5–13 per cent in Brazil, 6–16 per cent in Thailand and 15 per cent in Botswana and many of these men will also have women sexual partners (UNAIDS 2000).

As many as 90 per cent of women with HIV/AIDS have been infected through heterosexual intercourse and young women aged 15 to 24 are now the fastest-growing segment of the population contracting HIV/AIDS, with AIDS infection rates being three to five times higher in young women than in young men (Figure 4.4). In many countries, 70 per cent of women who die of AIDS are between 15 and 25 years of age (Smith 2002). 'The face of HIV has become that of young African women: seven of ten people living with the disease are in Sub Saharan Africa, and 58 per cent of infected Africans are female (9.2 million)' (Farley 2002: A1). These high levels of female infection are related to promiscuous male sexual behaviour and the prevalence of women sex workers in these areas. They are also influenced by women's lack of power, which makes it difficult for them to insist on the use of condoms to reduce disease transmission, even by husbands. Thus one of the main findings of reports presented at the international AIDS Conference held in Barcelona in July 2002 was that efforts to enhance the status of women in poor countries were central to reducing the spread of the AIDS pandemic (Bell 2002). In the Caribbean, adult women make up

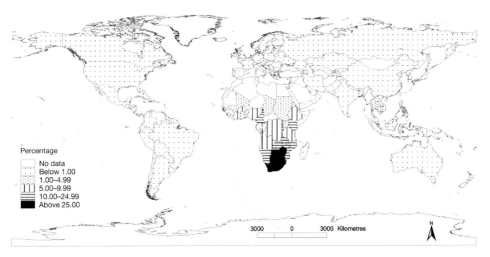

Figure 4.4 *HIV prevalence rate (%) in young women (ages 15–24); midpoint of high and low estimates, end of 2001.*
Source: UNAIDS (2002: 191, 195, 199)

less than half of those infected with AIDS where gender equality is greatest, as in Cuba, Jamaica and Trinidad and Tobago, but form a majority in the Dominican Republic (Figure 4.3). Some countries, such as Uganda and Brazil, have reduced the infection rate of HIV/AIDS through government-sponsored campaigns and widespread provision of condoms.

The World AIDS Campaign for 2002–3 is focusing on stigma and discrimination because it is felt that such attitudes associated with HIV and AIDS are the greatest barriers to preventing further infections, to providing adequate care, support and treatment and to alleviating impact (Aggleton and Parker 2002). This can be seen in the policies being implemented currently in Swaziland, where the percentage of young women infected with HIV/AIDS is 59.3. The Swazi King decreed a five-year sex ban on young women in September 2001. During this period, young women have to observe a prohibition on sex, including no shaking of hands with males and no wearing of pants; in addition, young women have to wear woollen tassels (symbolizing their untouchability) wherever they go for five years. The age group targeted has not been clearly stated, but women who are in relationships and older than 19 years will be expected to wear red and black tassels, and those who are still virgins will wear blue with yellow. This new intervention will be policed by traditional chiefs who still rule over much of Swazi society. Anyone who fails to observe the rules is fined the local equivalent of about US$150 or one cow. Phepsile Maseko, Coordinator of the National Youth Gender Caucus of Swaziland, stated in an interview that this edict demonstrates the government's unwillingness to deal effectively with the HIV/AIDS epidemic, as it does not address the need to care for those already infected with the virus nor prevent the spread of disease through improving public awareness and education programmes (AWID Resource Net 2001). It encourages male sexual behaviour and absolves them of responsibility in spreading the disease. It reinforces power imbalances within a patriarchal structure and explicitly blames women for HIV/AIDS. Furthermore, by identifying virgins it makes them a potential target for men looking for safe sex. Girls caught disobeying the law and who are unable to pay the fine have been thrown out of school (ibid.).

The rising number of women with HIV/AIDS has reversed earlier development gains. Female life expectancy in Botswana has fallen from 53 in 1970 to 47 in 2002 and this directly impacts the health and welfare of the family, as women do most of the work in farming and caring for the sick. Most HIV/AIDS victims live in rural areas, where poverty and poor health make a person more vulnerable to HIV infection and can shorten the incubation period for the virus and yet where there is least access to medical care. People with HIV/AIDS need to eat more protein than usual otherwise the immune system collapses. The common pattern of dependence on wild foods during famine is not an option for the sick (Carroll 2002a). Hunger breeds HIV/AIDS and HIV breeds hunger. It is projected that the nine hardest-hit African countries will lose between 13 and 26 per cent of their agricultural labour force between 1985 and 2020 to AIDS, with Namibia suffering the most (FAO 2002). HIV/AIDS undermines food security through its impact on a country's ability to import food by reducing per capita GDP growth by 1 per cent a year in Africa. Household food production declines because of labour shortages and because parents often die before passing on agricultural knowledge to their children, so AIDS and drought create a vicious cycle. In addition, a study in Zambia found that 68 per cent of rural orphans were not enrolled in school, compared to 48 per cent of non-orphans (ibid.).

In India, many HIV-infected women are choosing to terminate pregnancies because they know that their child would be rejected by society because of the stigma of the disease. Infected women are also ostracized but men with HIV/AIDS are much less likely to be stigmatized. Households cannot purchase food because of impoverishment due to loss of productive family members and of assets. Poverty, low levels of literacy and inaccessibility of rural areas make AIDS education and health services difficult to deliver.

Rural communities also bear a higher burden of the cost of HIV/AIDS because many urban dwellers and migrant labourers return to their villages when sick, further spreading the disease. Young sex workers in Bangkok often return to their natal villages in northern Thailand or Laos when they become infected and so introduce the disease to remote rural areas. The burden of caring for sick family members falls on women to a very large extent. Household expenditures rise to meet medical bills and funeral expenditures at the same time as the number of productive family members declines. Community organizations break down and are unable to provide assistance to orphans and the elderly victims of the epidemic. AIDS worsens existing gender-based differences in access to land and other resources in Africa. Some of the traditional mechanisms to ensure a widow's access to land contribute to the spread of AIDS, such as the custom that obliges a widow to marry her late husband's brother. In several countries women whose husbands have died of AIDS were forced to engage in commercial sex to survive because they had no legal rights to their husband's property.

Industry suffers also from absenteeism because of sickness and funerals, which increases costs of production. Most AIDS victims come from the best-educated working age group who are most mobile and can afford most sex partners. Consequently the loss of this group from AIDS has a greater effect on development than mere numbers might suggest. Many large South African firms are considering the cost effectiveness of paying for the drugs their HIV/AIDS workers need. Education also declines as teachers fall sick, health services cannot be maintained as medical staff die, and transport services also are affected since truck drivers are often the main sources of the infection, especially in India.

AIDS is having a destabilizing effect on society through the creation of millions of orphans and the loss of the best-educated people of working age. Although at the end of 2002 most of the estimated 42 million people with HIV/AIDS were in sub-Saharan Africa

(29.4 million), a new front is opening up in Asia, with a potential for a ten-year decline in average life expectancy by 2025 in India and China (Dyer *et al.* 2002). In 2002 almost one million people acquired HIV in Asia and the Pacific, a 10 per cent increase over 2001 (ibid.). It has been estimated that China may have 30 million and India 32 million AIDS sufferers by 2010 (ibid.). In China the disease has been spread through contaminated blood in transfusions, with as many as half the adults in some villages infected (ibid.), and by shared needle use by drug addicts. Both transmission methods affect men and women. Drug users in India near the border with Myanmar are also among the most affected, as are prostitutes and truck drivers. The growth of migrant labour in China is also likely to spread the disease and the pattern of conservative sexual behaviour, which has hitherto provided some protection, appears to be changing. Awareness campaigns can work to slow the spread of HIV/AIDS, especially among the young. Infection is falling in Thailand, Uganda and the Ukraine and even among young mothers in South Africa. In Addis Ababa, the capital of Ethiopia, the infection rate among young women has fallen by one-third (ibid.) but to achieve such success governments have to be willing to overcome political denial and social stigma. In India, China and South Africa politicians have been slow to confront the problem, and are only just beginning to set in motion treatment options to deal with this epidemic.

Gendered occupations and health

Gender-specific work exposes men and women to different environmental risks and thus to different causes of morbidity and mortality. Women's collection and use of water exposes them to waterborne diseases and parasites (Table 4.1). Spending time in poorly built houses increases the risk of Chagas disease (from insects living in mud walls) and smoke from cooking fires is also a health hazard (see Plate 4.2 and Chapter 5). Work in the fields exposes women to the dangers of pesticides and other chemicals which are increasingly used on crops. Women try to avoid applying chemicals as they realize the dangers they present not only to themselves, but also to their unborn or nursing children. For women there is an increased risk of late births, miscarriage and stillbirths and some herbicides may interfere with women's oestrogen levels, altering normal menstrual cycles (Ransome 2001). The presence of pesticides in breast milk is a concern. For example, it has been estimated that in

Table 4.1 *Work-related health risks for women*

Health problem	Gender-specific related cause
Schistosomiasis (bilharzia)	Washing themselves, children and clothes in streams
Dengue; Chagas' disease, arsenicosis; leishmaniasis	Domestic roles in the house
Burns; back pain	Cooking on open fires or stoves
Respiratory infections; coughs; lung cancer; detrimental effects on foetal growth and children	Cooking in poorly ventilated structures using biomass fuel sources.
Fatigue and muscle pains in legs, shoulders and hips; prolapsed uterus; miscarriage; stillbirth	Carrying heavy loads of water and fuelwood or crops from fields.
Headache; broken bones	
Malaria; filariasis; chronic back pain; leg problems; cuts	Farmwork involving constant bending, e.g. weeding, transplanting, threshing, post-harvest processing
Exposure to toxic chemicals with effects also on unborn and breastfed infants	Cash crop production: working in sprayed fields and in greenhouses without protective clothing, agroprocessing
Eye problems; exposure to toxic chemicals	Electronic assembly line work
Urinary tract infections; back leg and foot pains; accidents.	Factory work without adequate frequency of bathroom breaks and with long periods of standing
STDs and HIV/AIDS	Sex work

Source: adapted in part from Oxaal and Cook (1998).

Plate 4.2
Mexico: a woman making tortillas on an open fire inside a house in a Mayan village in Quintana Roo.
Source: author

Delhi, India, the average infant receives 12 times the acceptable level of DDT, an extremely hazardous pesticide long banned in developed countries (ibid.). Labels on these chemicals are often only in English and so may be of no help to most farmers, and protective clothing is rarely worn. In fact male farmers may think it 'macho' not to bother with such protection. Women are usually the ones who wash the clothes used while applying chemicals, so are becoming exposed. Farm chemicals are often stored in the home, putting all members of the household at risk, while empty chemical containers may be used for storing food or water.

Many women are now taking up work in agro-processing and manufacturing. In agro-processing, such as picking and post-harvest preparation of flowers or bananas, they may also be exposed to chemicals. In processing cashew nuts, women workers are rarely provided with protective clothing and so may get burns from the acid in the shell of the nut. In garment factories pressure to work ever faster may result in accidents with machinery. Electronics manufacturing may lead to eye strain and exposure to chemicals. In most of these jobs the hours are long with very limited breaks and women are on their feet much of the time (Table 4.1). Deaths of workers in electronics firms in northern Thailand were thought to be due to lead poisoning (Glassman 2001). Companies are competing for contracts and so constantly increase the pressure on workers to be more productive, making for high stress levels. Working late means young women travelling between home and work in the dark, increasing the likelihood of attack. In many societies young women factory workers, having left the protection of their parental home, are seen as loose and vulnerable (Buang 1993; Navarro 2002).

Violence

Violence is now seen as a health problem and protection against violence as a human right. Overall, violence, whether self-inflicted, interpersonal or collective, is among the leading causes of death among people aged 15 to 44 years (Krug et al. 2002). Violence can be prevented and efforts to do so are being implemented at the local, national and international level. The World Health Organization defines violence as 'The intentional use of physical force or power, threatened or actual, against oneself, another person, or against a group or community, that either results in or has a high likelihood of resulting in injury, death, psychological harm, maldevelopment or deprivation' (ibid.). This definition associates intentionality with violence and includes acts, such as neglect and psychological abuse, resulting from a power relationship. It also divides violence into self-directed, such as suicide or self-abuse, interpersonal, including family and intimate partner violence and community violence, and collective, which is subdivided into social, economic and political violence (Krug *et al.* 2002). However, in practice the dividing lines between the various types of violence are not always clear.

In 2000 some 91 per cent of violent deaths in the world occurred in low- and middle-income countries and almost half of such deaths were suicides (ibid.). Homicide and suicide rates are much higher for men than for women. In Africa and the Americas homicide rates are nearly three times suicide rates, but in Europe and South-east Asia suicide rates are more than double homicide rates (ibid.). In East Asia and the western Pacific suicide rates are nearly six times homicide rates (ibid.) and, unusually, in rural China more women than men commit suicide. There are also other variations within regions and between ethnic groups, rich and poor, and rural and urban populations (ibid.).

Societies with high levels of inequality and experiencing rapid social change often have an increasing level of interpersonal violence. In a study of poor households in 23 transitional and developing countries only 9 per cent of respondents reported that domestic violence against women was rare; 30 per cent felt there was a decrease and 21 per cent reported an increase, while the remainder reported little change (Narayan *et al.* 2000a). The countries of Eastern Europe and Central Asia reported the largest increase (32 per cent), while both Latin America and the Caribbean and Asia reported decreases of 44 per cent and 41 per cent respectively (ibid.).

Following the focus on violence against women at the Vienna Conference on Human Rights in 1993, the United Nations Commission on Human Rights appointed a special rapporteur on violence against women (United Nations 1996). This preparatory work led the Beijing Platform for Action (BPFA), which came out of the fourth World Conference on Women in 1995, to identify violence against women as one of the 12 critical areas of concern, declaring it 'an obstacle to the achievement of equality, development, and peace' (ibid.: 676 D112). The BPFA recognized that 'The low social status of women can be both a cause and a consequence of violence against women' (ibid.: 676 D112). It defined violence against women as meaning 'any act of gender-based violence that results in, or is likely to result in, physical, sexual, or psychological harm or suffering to women, including threats of such acts, coercion or arbitrary deprivations of liberty, whether occurring in public or private life' (ibid.: 676 D113). It further points out that:

> Violence against women throughout the life cycle derives essentially from cultural patterns, in particular the harmful effects of certain traditional or customary practices and all acts of extremism linked to race, sex, language or religion that perpetuate the lower status accorded to women in family, the workplace, the community and the society.
>
> (ibid.: 677 D118)

In many cases, violence against women and girls occurs within the home, where violence is often tolerated. The fact that women are often emotionally involved with and economically dependent on those who victimize them has major implications (see Box 4.1). The neglect, physical and sexual abuse, and rape of girl-children and women by family members and other members of the household, as well as incidences of spousal and non-spousal abuse, often go unreported and are thus difficult to detect.

Table 4.2 *Types of violence*

Type	Examples
Social	Killing or rape to protect family or group 'honour', especially in societies following a narrow view of Islam
	High rates of suicide as among rural women in China because of their low social status
	Disfigurement by throwing acid at young women who reject a suitor
	Female infanticide because of son preference
	Female genital cutting
Economic	Dowry deaths where the husband's family considers the dowry inadequate
	Trafficking in women and sale of poor women and young boys into prostitution and as slaves
	Female infanticide in order to save the natal family the costs of raising a girl
	Backlash against women who receive microcredit loans not available to men or who are able to get jobs in female-dominated industries
Political	Rape as ethnic cleansing in war
	Sexual slavery
	Forced adoptions

Box 4.1

Family violence in Tajikistan: the tale of Fotima and Ahmed

Fotima, 35, her four children and her husband, Ahmed, live with the latter's parents and elder brother in Kurgan Teppa in the Central Asian Republic of Tajikistan, a few miles north of the Afghan border. Life is hard, since a civil war (1992–7) exacerbated the economic collapse of the post-Soviet period. Ahmed is not a very forceful or effective person and does not manage well in the new environment, where it is important to be able to hustle to make a living. By importing goods from abroad his elder brother earns enough to support not only his wife and children but also his parents, while Ahmed's income is so meagre that Fotima has to supplement it by petty trading. As a result Ahmed's father is very scornful of him, continually taunting him and calling him names that suggest he is less than a man. During Soviet times, when Ahmed had been a teacher with an adequate salary, he and Fotima got on well. Now their relationship is becoming increasingly strained. When his father taunts him Ahmed cannot answer back so instead he takes his frustration out on his wife. He hasn't yet hit her, although she senses that the moment is not far off, but he screams and shouts at her until she can hardly bear it. Afterwards she finds herself taking out her own frustrations on her children, whom she frequently yells at and slaps round the head. Lately the situation has worsened and Fotima at times finds herself hitting the children so hard she is frightened of what may happen.

She approached the Kurgan Teppa Women's Centre, which caters for women with problems, especially those related to domestic violence, and begged the guidance counsellors there to help her stop abusing her children. After listening to Fotima's story, the Centre's counsellors started to work with her to find other ways of dealing with her emotional stress. They have asked Fotima to invite Ahmed to visit them also but she is too scared to let him know she has told their story to outsiders. She prefers to continue to put up with the situation, while trying to follow the counsellors' advice on how to work through her pain in such a way as to protect her children.

Such tales as that of Fotima and Ahmed are only too common in Tajikistan today, where domestic violence of all kinds is rife. Increasing numbers of cases come daily to the Women's Centre, run by the Tajik NGO Ghamkhori, with which I have collaborated since its inception in 1997. Besides the Centre, Ghamkhori has other projects that support rural communities to increase their control over their lives and environment, and reduce the rate of overall family violence. Using a specially elaborated participatory methodology the projects involve as many community members as possible, from religious and secular leaders, through schoolteachers and medical professionals, down to teenagers, and has a high success rate in improving family relationships and thus decreasing violence in the home.

Source: Colette Harris, Women in International Development, Office of International Research, Education and Development, Virginia Tech University, September 2002.

In 48 surveys from around the world, between 10 per cent and 69 per cent of women reported being physically assaulted by an intimate male partner at some time in their lives (WHO 2002). A survey in Armenia revealed that 46 per cent of the participants were victims of violence at home and UNICEF estimates that 76 per cent of all violence against women in Guatemala occurs in the home (WOAT/OMCT 2001). Even when such violence is reported, there is often a failure to protect victims or punish perpetrators (United Nations 1996: 677 D117). By June 2000, at the Beijing +5 Review, strategies for reducing such violence had been' laid out. These strategies focused on the role of men, and on men's and boys gender roles and actions to break the cycle of violence (Gonzalez 2001). Masculinities have increasingly become a research focus in attempting to help men to understand and control their aggressive tendencies towards women. Three types of violence against women have been recognized: social, economic and political (see Table 4.2).

Violence and socio-cultural links

In many societies the honour of the family depends on protecting the virginity of their daughters and preventing women and girls from bringing shame on the family through their public behaviour (Box 4.2). Women are expected to cover themselves and to not be seen by men from outside the family unless they have a male family member or sometimes an older female family member to protect them. Such attitudes were taken to the extreme in Afghanistan under the Taliban but milder versions are found in many countries. Under sharia law, as in Pakistan or northern Nigeria, a woman who is raped can be stoned to death. In Alexandria, Egypt, it was found that 47 per cent of murdered women were killed by relatives after they had been raped (Krug et al. 2002). In Sri Lanka, many Muslim families do not allow women to work outside the home in their own country, but do permit them to work as domestics in the Middle East because, although they are often treated with great cruelty, this is not visible to their own society and they are working in the land of Allah (Ismail 1999b). In many traditional societies wife-beating is seen as culturally and religiously justified. Women who are abused are more likely to suffer from depression, to attempt suicide, to earn less than other women and to experience a pregnancy loss and the death of children in infancy or early childhood (Krug et al. 2002).

Box 4.2

Honour crimes: a conflict between modern lifestyles and rural customs

Sait Kina saw his 13-year-old daughter as bringing nothing but dishonour to his family. She talked to boys on the street and she ran away from home. In spring 2001, when she tried to run away yet again, Kina grabbed a kitchen knife and an axe and stabbed and beat the girl to death in the bathroom of the family's Istanbul apartment. He then commanded one of his daughters-in-law to clean up the blood and mess. When his two sons came home from work 14 hours later, he ordered them to dispose of the corpse. 'I fulfilled my duty,' Kina told police after he was arrested. 'We killed her for going out with boys.' This teenager's behaviour was seen as bringing shame on the family so it had to be dealt with in the time-honoured way.

Dilber Kina's death was an 'honour killing', a practice occurring with increasing frequency in cities across Turkey and in other developing countries, where large-scale migration to urban areas has left families struggling to reconcile modern lifestyles and liberties with long-standing rural customs. Mounting social pressures have led to an alarming increase in murders (at least 200 a year in Turkey), beatings and other violence within families, as well as suicides among urban and rural girls and women.

Where a woman's honour is a family's only measurable commodity in an impoverished community, preservation of family status takes precedence over an individual's human rights. When a woman in the family is considered to have besmirched the family honour, male family members gather to vote on her death and to decide who will carry out the killing. The chosen assassin is usually someone under the age of 18 because his youth will allow him to be treated more leniently under the law. In Turkey the killing of a family member is a capital crime punishable by death or life in prison. But if a judge rules there was provocation for the killing, such as a question of honour, the penalty can be reduced. If the defendant is a minor and behaves well during the trial and detention in jail, the penalty is frequently cut to two years or less.

It is hard to quantify the global number of women and girls killed by family members, for the 'dishonour' of being raped. Most of these killings occur in predominantly Muslim countries, such as Egypt, Bangladesh and Pakistan, but they are also taking place in rich countries, such as Sweden and Britain, where there are migrant Muslim communities with educated young women who try to behave like their non-Muslim peers. A Somalian refugee was forced to flee the Netherlands in 2002 after getting death threats from fellow Muslims because she criticized Islamic attitudes to women, but in 2003 she was elected to the Dutch Parliament on a platform of emancipation for immigrant women.

Sources: adapted in part from Moore (2001) and Simons (2003).

Violence linked to economics

The young brides killed or severely burned in India by the husband's family are seen as victims of the ancient social custom of dowries. It is thought that they are doused with kerosene and set alight for failing to satisfy the demands of their husband's families for gold, cash and consumer goods that come as part of the marriage arrangement between families. Dowry deaths and suicides rose to 6,975 in 1998 from 4,648 in 1995 according to official figures, but researchers now believe that only about one-quarter of these are related to dowry harassment (Dugger 2000). Many women say that domestic violence is a normal part of married life in India and occurs most commonly for neglecting housekeeping or child-rearing duties, showing disrespect to in-laws, going out without a husband's permission, or arousing his suspicions of infidelity. However, there are also accidental injuries caused by the use of cheap, pump-action kerosene stoves, that are often shoddily made and lack even the most basic safety features but which are popular among the urban poor. Official figures show that, in 1998, 7,165 people died in stove accidents. Of this total 1,280 were men, suggesting that not all stove deaths are the result of hidden domestic violence against wives (ibid.). The use of fire as a weapon in domestic violence in India is simply expedient, as kerosene is cheap and usually at hand in the house.

It has also been argued that the violence against women factory workers is related to economics. According to the International Labour Organization (1999) women provide about 80 per cent of the workers in over 200 export-processing zones located in 50 countries. Young women working in these factories are transgressing social norms in many countries. This allows factory foremen to justify sexual harassment of these workers. In addition, these workers may be earning more than local men, which creates jealousy and undermines patriarchal household relations. Thus these women are ascribed particular identities laden with assumptions about their worth, value and respectability. Construction of these particular and subordinated identities facilitates and legitimates physical violence and also constitutes a form of violence in itself. Such representational violence may lie behind the disappearance of over 450 women and another 284 who have been found murdered between 1993 and 2002 in Cuidad Juárez on the Mexico/USA border, where there are currently 396 *maquiladoras* (export-oriented assembly factories) employing over 220,000 people, mostly women (Garwood 2002; Navarro 2002). The violence has now spread to women students and store clerks and most victims are between the ages of 15 and 25. No adequate response to this tragic violence has been made by Mexican politicians or law enforcement officers and Mexican women are leading public protests against what is seen as an entrenched culture of official impunity.

Trafficking

Trafficking in women and children is a lucrative business and, unlike arms and drugs, the victims can be sold many times (see Box 4.3). The United Nations in 1997 estimated that the trade was worth US$7 billion annually and so more lucrative than the trade in illicit weapons (USAID 1999). The number of women trafficked annually is somewhere between 700,000 and four million, with 50,000 coming to the USA (AWID 2002a). A survey in Armenia revealed that 32 per cent of Armenian women have a family member or close friend who has been trafficked (WOAT/OMCT 2001). Trafficking occurs both within borders and between countries. Countries of origin, destination and transit are intertwined. In South and South-east Asia the main countries of origin are Thailand, Bangladesh, Vietnam, Cambodia, China, the Philippines, Myanmar and Nepal. The main destinations and transit countries are Thailand, Malaysia, Japan, India and Pakistan (Wennerholm 2002). The purpose of this traffic is generally for commercial sex work, but women are also trafficked for domestic service and other forms of bonded labour and for marriage. Boys are trafficked for commercial sex work in Sri Lanka and from Bangladesh to the Persian Gulf to work as camel jockeys (Sengupta 2002). Latin American countries have a long tradition of trafficking in women and their destination countries are mainly in Western Europe, Japan and the USA. Since the fall of the Iron Curtain, trafficking from Eastern and Central European countries and former Soviet Central Asian republics to Western Europe has increased enormously. The war in the Balkans has encouraged the trafficking of women and children into Mediterranean countries, a trade supposedly dominated by the Albanian mafia. Women are trafficked to the United States for commercial sex, domestic service, bonded labour, illicit adoptions and as mail-order brides. In some parts of India, Nepal and Ghana sex slavery is linked to religious groups (Herzfeld 2002).

The victims of trafficking rarely know what to expect at their destination. They may be sold to traffickers by poor parents who have

Box 4.3

Husbands trafficking in wives in Bangladesh

Rural women in Bangladesh want to get married and try to avoid bringing shame on their natal families by being rejected by their husbands. These attitudes were exploited by traffickers in Jhikargacha, a small village in Jessore district, west-south-west of Dhaka, near the border with India. This area is well connected by bus and train to Calcutta, where there is large-scale prostitution and selling of women, and agents to take women to Bombay and Pakistan. Many Bangladeshi men are employed in India and they come back to their home villages to get married. After marriage, the wife accompanies her new husband to his place of work and then disappears. The husbands often return to Bangladesh to remarry. Since Islam allows men to have more than one wife if they can support them, poor parents agree to the marriage, thinking it will ensure a prosperous future for their daughters. Border guards cannot stop a woman travelling with her husband, and it is only after several incidents that parents become suspicious. By then their daughters have disappeared, probably sold in India, Pakistan or the Middle East.

Jahanara, of Jhikargacha village, was married, but because of non-payment of her dowry she was divorced and returned to her parent's home with her newborn son. She was married again to Hossain Ali. Ali had married a village girl, Bella, two years earlier and after two months in the village had disappeared with Bella and her younger sister Pachi. On his return to the village he told Bella's parents that both girls had got good jobs in Bombay. After some months he sent a proposal to marry Jahanara and her parents agreed, as it is not easy for divorced women to remarry. Soon after the marriage he persuaded Jahanara to go with him to Bombay on condition she left her son behind. She agreed, but her mother tried to dissuade her. However, she left surreptitiously during the night with Ali, leaving behind her infant child. The child suffered without his mother's care and died a few months later. After this incident it became obvious to the villagers that Hossain Ali was a procurer and had used marriage as a ruse to obtain women to be sold to pimps and brothel owners in Bombay.

Source: adapted from Shamin (1992).

been told their children will be given good jobs. In a strange country they are vulnerable, as they may not speak the language, and they are tightly controlled by their pimps. They are often exposed to violent treatment, become infected with STDs and HIV/AIDS and rarely have access to medical services. If they escape they are treated as illegal immigrants and deported back to the poverty they had hoped to leave behind. Occasionally they are able to send money back to their families and may become an important economic source. If they return home those trafficked as children may not recognize their parents nor remember their native language (Sengupta 2002). The women may be unable to return to their home villages because they feel stigmatized and they may also be infected with HIV/AIDS. The cost of re-socializing these returnees, educating children and retraining women is high. Women may be given training in handicrafts or even as taxi drivers, but they are rarely able to make much money and many become traffickers themselves (see Box 8.3).

In December 2000 over 80 countries signed the Protocol to Suppress, Prevent and Punish Trafficking in Persons, Especially Women and Children (The Trafficking Protocol) in Palermo, Italy. This brought up to date the 1949 Convention for the Suppression of the Trafficking in Persons and Exploitation of the Prostitution of Others (the Trafficking Convention). The definition of trafficking in the United Nations Protocol of 2000 says in part:

> 'Trafficking in persons' shall mean the recruitment, transportation, transfer, harbouring, or receipt of persons, by means of the threat or use of force or other forms of coercion, of abduction, of fraud, of deception, of the abuse of power or of a position of vulnerability or of the giving or receiving of payments or benefits to achieve the consent of a person having control over another person, for the purpose of exploitation.
>
> (quoted in Williams and Masika 2002)

It does not make all prostitution illegal, but it does require signatories to the Protocol to extend assistance and human rights protection to the victims of trafficking (Williams and Masika 2002). From 2003 the United States will penalize countries that make no effort to halt the practice (AWID 2002a). Penalties could include votes against loans from the International Monetary Fund and the World Bank. Of 89 countries examined recently by the United States, it was found that 19 were not doing enough to stop trafficking across international borders, compared to 23 the previous year.

However, critics have pointed out that compliance may merely mean making legal migration more difficult. In Bangladesh, over the last 15 years, 500,000 females have been lured out of the country with promises of work or marriage, but many were simply sold into brothels (Box 4.3). In response Bangladesh effectively bars women, except for skilled professionals, though not men, from working overseas legally, but poor women still go illegally (Sengupta 2002). There is now a trafficking law with stiff penalties but only low-level operators, not the kingpins, who are wealthy enough to pay bribes, have been arrested. A public education campaign against trafficking is under way, which includes special training for police and border guards, but it is doubtful if this can stop such a lucrative business as long as the demand is there.

Political violence

Women may be active participants in war or they may be victims. Women as fighters learn new skills and are empowered but are usually pushed back into subordinate positions when peace comes. This rejection of wartime social transformations of roles and identities is less true where women are increasingly playing a role in government. In the post-cold war period there have been 49 conflicts and 90 per cent of those killed in these conflicts have been non-combatants (Saferworld 2002). Ethiopia's Minister of State in charge of Women's Affairs in the Prime Minister's Office noted that the victims of war were also women and children, yet: 'Although the involvement of women is considered to be vital for ensuring sustainable peace, women have so far been marginalized and do not participate fully . . . in conflict prevention and resolution, as well as in peace initiatives' (Abasiya 2002). The Secretary General of the United Nations said: 'Women play an active role in informal peace processes, serving as peace activists, including by organizing and lobbying for disarmament and striving to bring about reconciliation and security before, during and after conflicts' (Annan 2002). He urged that gender units should be set up within peacekeeping operations, that women should be fully involved in peace negotiations and that 'sustainable peace will not be achieved without the full and equal participation of women and men' (ibid.). Women are particularly at risk of human rights abuses in conflict situations because of their lack of status in most societies. Women are also disproportionately affected by the lack of basic services endemic to conflict and displacement, such as adequate medical attention, nutrition, sanitation and shelter (Rehn and Sirleaf 2002). Women's involvement in post-conflict activities have been shown to be important, as in Albania, where a UNDP weapons-for-development programme collection was highly successful due entirely to the participation of women (United Nations 2002). Since 2000 women have been actively involved in peace negotiations in Burundi, Afghanistan and in the Democratic Republic of Congo (ibid.).

Rape as a weapon of ethnic cleansing, as it was used in Rwanda and in Bosnia-Herzogovina, has now been recognized by the United

Nations as a 'crime against humanity'. This has encouraged the delayed recognition of Japan's rapes, sexual slavery and abduction of women from all the colonized and occupied countries during the Second World War to be 'comfort women' for the Japanese Imperial Army, as war crimes. A War Crimes Tribunal was held in Tokyo in December 2000 to draw attention to these crimes and to try and get reparations for the now elderly victims. Sexual slavery was also used during more recent wars in Angola and Mozambique. Under the military dictatorship in Argentina men and women were imprisoned arbitrarily and if women gave birth in prison the babies were given anonymously to government supporters. Most of the families of these prisoners have never found these children who were born in prison and grandmothers continue to publicly protest (Radcliffe and Westwood 1993).

Women forced into refugee camps have to learn to cope alone. They may suffer violence as in many of the camps in Africa, or they may learn new skills, such as literacy and technical training, which empower them in the post-conflict situation, as in El Salvador. In the post-conflict situation many women find themselves as heads of households, but without the resources to support their families, as in Rwanda, where women head half the families but widows have no rights to their husband's land. In addition, many people suffer from post-traumatic stress, especially where physical and psychological torture has been used (Leslie 2001). They may also need help in adjusting to the new social and political situation, where the nation's infrastructure has been destroyed and there are few jobs.

Changing attitudes

In the last decade much has been done to improve the health of poor women. The BPFA stated that 'women have the right to the enjoyment of the highest attainable standard of physical and mental health' (United Nations 1996: 667 S89). It also recognized that 'women have different and unequal access to and use of basic health resources' (ibid.: 667 S90). Strategic objective C1 of the BPFA was to '[i]ncrease women's access throughout the life cycle to appropriate, affordable and quality health care, information and related services' (ibid.: 670).

This new focus on health and protection from violence as human rights led to a rapid expansion of preventive health care and community-based programmes, usually run by NGOs, with a new focus on including men (Smitasiri and Dhanamitta 1999; Sanez *et al.* 1998) (Box 4.1). Educating men in their role in women and children's health has apparently been remarkably successful in changing social attitudes and gender roles in countries as different as India and Costa Rica. The Philippine Plan for Gender-Responsive Development (PPGD, 1995–2025) is a model plan for other governments in the region. This plan has several projects, with men working on topics as varied as male responsibility in breastfeeding children, men and the fight against violence against women and a grassroots project working with men in promoting reproductive rights and health (Women in Action, 2001). Even in highly patriarchal societies some men can remain marginal to the dominant order of patriarchy and be open to change. Such change will be slow and incremental and there is no clear blueprint for gender-inclusive public policy.

Learning outcomes

- Age and poverty influence health.
- HIV/AIDS is gendered and is most commonly transmitted through heterosexual sex.
- Occupation affects health.
- Violence is of many types and part of the solution involves changing men's attitudes.

Discussion questions

1 How do socio-cultural ideas effect the gendered infection rate of HIV/AIDS?

2 What is the impact on the source nations of the trafficking in women and children?

3 What social attitudes need to be changed to improve women's health?

Further reading

Dyck, Isabel, Nancy Davis Lewis and Sara McLafferty (eds) (2001) *Geographies of Women's Health*, London and New York: Routledge. An edited collection focusing on women's health, with case study chapters on Africa, Thailand, India and Papua New Guinea and an introduction which considers the impact of globalization on women's health.

Krug, Étienne G., Linda L. Dahlberg, James A. Mercy, Anthony B. Zwi and Rafael Lozano (eds) (2002) *World Report on Violence and Health*, Geneva: World Health Organization. Provides some of the most recent worldwide data on violence and health.

Moser, Caroline O. N. and Fiona C. Clark (2001) 'Gender, conflict, and building sustainable peace: recent lessons from Latin America', *Gender and Development*, 9 (3): 29–39. A short article discussing the gendered impact of conflict in Latin America.

Salomon, Joshua A. and Christopher J. L. Murray (2002) 'The epidemiologic transition revisited: compositional models for causes of death by age and sex', *Population and Development Review* 28 (2): 205–28. An important paper providing current views on gender and age aspects of health.

Websites

www.who.int/home-page   The World Health Organization (WHO). The main objective of the WHO is the attainment by all peoples of the highest possible level of health. The site gives insight into WHO programmes and activities.

www.who.int/frh-whd/index.html   The official website of the Women's Health Department Homepage of the WHO. This site contains information on women and HIV/AIDS, FGM, reproductive health, and violence against women.

www.ishc.org   The International Women's Health Coalition's website.

www.who.int/ageing/index.html   World Health Organization's Ageing and Health Programme.

www.trafficked-women.org   Coalition to Abolish Slavery and Trafficking (CAST) is an alliance of non-profit service providers, grass-roots advocacy groups and activists dedicated to providing services and human rights advocacy to victims of contemporary slavery.

www.antitrafficking.org   CHANGE aims to promote and protect women's human rights worldwide. It is undertaking an anti-trafficking programme.

www.antislavery.org   Anti-Slavery International was set up in 1839 with the objective of ending slavery worldwide. It publishes information on slavery and promotes laws to protect those exploited by such practices.

www.inet.co.th/gaatw   Global Alliance against Trafficking in Women (GAATW) was founded in 1994 in Thailand. It facilitates and coordinates work on trafficking in persons and women's labour migration throughout the world.

www.unaids.org   UNAIDS, the United Nations AIDS programme, runs campaigns annually to make the problem more widely understood and to encourage the search for solutions.

⑤ Gender and environment

Learning objectives

When you have finished reading this chapter, you should be able to:

- understand ecofeminism and its various alternatives
- be aware of the factors influencing gender differences in environmental perception
- realize how natural resource use is gendered
- appreciate the gendered impact of pollution.

The final decade of the twentieth century witnessed increasing interest in the analysis of women/environment interaction and the gendered impact of environmental policies. Meetings such as the Global Assembly of Women and the Environment in Miami in November 1991 and the Global Forum, held in Rio de Janeiro in June 1992, aimed at both development activists and popular audiences. The Women's Action Agenda 21 which resulted from these meetings is a call for feminist collaboration in environmental action and goes far beyond the official UNCED position on women and the environment contained in Chapter 24 of 'Agenda 21', the global action plan adopted in Rio de Janeiro. Chapter 24 is, however, a major step forward in attitudes on the part of signatory governments, although limited by the structural inertia of official policy formulation and a resource-based approach to sustainable development.

The Rio Earth Summit led to the founding of the Women's Environment and Development Organization (WEDO), an international network of women's organizations, and of the Women in Europe for a Common Future (WECF), a pan-European women's environmental organization. After Rio increasing numbers of women became involved in issues of environmental sustainability, biodiversity, climate change and protection of natural resources. These women bring a gender analysis and a human rights framework to these issues (Khosla 2002a).

Most governments have not lived up to their commitments made at the Rio Earth Summit. Recognizing this, the World Summit on Sustainable Development (WSSD), held in Johannesburg in 2002, focused on implementation. The ideas of the Rio Agenda 21, the Millennium Development Goals adopted by most countries in 2000, and the commitments made at the World Trade Organization Doha Ministerial meeting of 2001 and at the Monterey March 2002 Conference on Financing for Development were all rolled into one Plan of Implementation. The WSSD decided that the five key areas in which action should be taken in order to relieve poverty were water, energy, health, agriculture and biodiversity (Percival 2002). In the decade between the Rio and Johannesburg conferences, 95 per cent of the global population growth had occurred in those countries classified by the United Nations as less developed. These countries' populations grew by 18 per cent over this period, and those of the 48 least developed, mostly African, countries by 29 per cent, compared to a 3 per cent growth rate in developed countries.

This growing population in the poorest countries, increased recognition of the role of transnational corporations in globalization and underfunding of the United Nations have led to a new emphasis on public/private partnerships. The United Nations proposed in Johannesburg that private partnerships should be formed between any combination of civil society organizations, governments, UN agencies and the private sector to implement projects for sustainable development. Women's groups at the Johannesburg Summit asked about the accountability of such partnerships and suggested that they might be a cover to absolve national governments from their responsibilities for implementing the Rio Earth Summit commitments. Women's voices were a strong presence at the Johannesburg Summit and made it clear that there will be no sustainable development without gendered analysis.

Women in many parts of the world are involved in grassroots environmentalist activism (Rocheleau et al. 1996). They have also fought against local toxic waste issues (Miller *et al.* 1996; Bru-Bistuer 1996) and against destruction of forests (Wastl-Walter 1996; Campbell *et al.*, 1996; Sarin 1995; Nesmith and Radcliffe 1993). Many have seen these activities and the associated high-profile social movements as proof that women's natural closeness to nature makes them more aware of environmental issues than men resulting in the ideology of ecofeminism (Shiva 1989; Mies and Shiva 1993). However, although many of these grassroots movements are fundamentally humanitarian, pluralistic and activist, women's organizations are neither inherently altruistic nor environmentalist. The Indian Chipko movement, often seen as an attempt to protect forest resources by local women, has recently been interpreted as being not an example of women's links with the ecological needs of their region, but in reality part of a broader current of peasant protest.

What little information exists on gender differences in environmental perception and values tends to show that, although women may be more concerned about environmental issues than men, they are less politically active on these issues (McStay and Dunlap 1983). Case studies from many countries reveal that differences on environmental priorities between the genders tend to be modest. Studies of rural communities often show diverse links between environmental attributes and gender (Leach *et al.* 1995). As a result, no firm conclusions can be drawn as to gender differences in general environmental perception.

Much of the contemporary work on examining gender differences in environmental perception is driven by an interest in understanding women's role in the environmental movement and in harnessing them as 'managers' of the environment. It is often asserted that women's relationship with the environment is 'special' and that women are more motivated than men to work for the enhancement of the sustainability of the environment. This has encouraged development agencies to assume a synergy between women and environment when allocating aid, with the result that 'there are serious risks of simply adding environment' to the already long list of women's caring roles, instrumentalizing women as a source of cheap or unrewarded labour' (ibid.: 7). A gender-based approach to environmental issues, rather than a narrow focus on women's environmental roles, can enable separate, complementary and conflicting interests to be identified in ways that should lead to improvements in the sustainability and equity of environmental policy.

Ecofeminism

'Ecofeminism' has been defined as 'a movement that makes
connections between environmentalisms and feminisms: more
precisely, it articulates the theory that the ideologies that authorize
injustices based on gender, race and class are related to the
ideologies that sanction the exploitation and degradation of the
environment' (Sturgeon 1997: 25). In this definition, Noel Sturgeon
provides a very broad and apparently innocuous definition of
ecofeminism, yet most recent work on gender and the environment
(Agarwal 1992; Leach *et al.* 1995; Rocheleau *et al.* 1996; Sachs
1997; Seager 1997) insists on locating itself outside ecofeminism.
Silvey (1998) suggests that political activists have been more willing
to identify with ecofeminism than have academics, perhaps because
research labelled ecofeminist was actively excluded from many
academic agendas (Gaard 1994; Sturgeon 1997), but she argues that
linking one's research to ecofeminism can contribute to feminist
environmental and political goals. Ecofeminism may be broken down
into four different types: liberal, cultural, social and socialist
(Merchant 1992). All ecofeminists share an environmental 'ethic of
care' based on women's biology, labour or social position but their
strategies for change differ. Liberal ecofeminists tend to work within
existing structures of governance by changing laws and regulations
relating to women and the environment. Cultural ecofeminists
criticize patriarchy and emphasize the symbolic and biological links
between women and nature. Social and socialist ecofeminists focus
on social justice issues and analyse the ways in which both
patriarchy and capitalism contribute to men's domination of women
and nature. Sachs (1997), however, emphasizes the similarities
showing that both ecofeminists and their critics focus on three major
issues: (1) women's relationships with nature; (2) the connections
between the domination of women and the domination of nature; and
(3) the role of women in solving ecological problems.

The separation between nature and culture is paralleled by other
dualisms of female/male, body/mind and emotion/reason. Western
philosophy links women with nature, body and emotion, while men
are associated with culture, mind and reason. Women are considered
to be more environmentally sensitive than men because of their
traditional caring and nurturing role. It is suggested that the
preconceived similarities of passivity and life-giving qualities
between women and nature make both equally vulnerable to male
domination (Merchant 1992).

Ecofeminism consists of several strands relating to the connections between women and nature. Two major tenets are, first, the co-domination of women and nature (Plumwood 1993; Warren 1990); second, Shiva (1989) extends this to blame Western science and colonial development policies for the negative impact of economic development on both the environment and on women's lives in the global South. Cultural ecofeminists emphasize the importance of biology in bringing women closer to nature, arguing that the female biological processes of pregnancy and childbirth are the source of women's power and ecological activism. Such views have been critiqued as essentialist, universalist, reductionist and as having a focus on personal spirituality. These criticisms are briefly reviewed in the following section.

Many social scientists, especially those working on development issues, find ecofeminist views based in biology unhelpful. Dianne Rocheleau and colleagues state that there are '*real* not imagined, gender differences in experiences of, responsibility for, and interests in "nature" and environments . . . and these differences are not rooted in biology *per se*' (1996: 3). They prefer a feminist political ecology approach, which brings a feminist perspective to political ecology and 'treats gender as a critical variable in shaping resource access and control, interacting with class, caste, race, ethnicity to shape processes of ecological change' (ibid.: 4). Agarwal (1992) takes the critique of ecofeminism's universalist and anti-materialist views further, pointing out that ecofeminism fails to take into account not only differences of class and race but also of occupation and geographical context. She, like Rocheleau *et al.*, insists that an understanding of the connections between people and the environment requires a critique grounded in the realities of men's and women's lives. She proposes an alternative theoretical position of feminist environmentalism based on regional patterns of gendered differences in divisions of labour, property ownership and power (Agarwal 1997a). Sachs (1997) also stresses the importance of difference, suggesting that a postmodern emphasis on local knowledge rather than universal truths is especially useful for exploring women's understanding of the environment. Others critique the focus on women alone as being too narrow, as it makes men invisible (Braidotti *et al.* 1994; Leach *et al.* 1995), and instead emphasize the importance of understanding processes of resource use and their structuring by gender relations. Thus they stress a move from a Women, Environment and Development (WED) approach to one of Gender, Environment and Development (GED).

Such postmodern approaches have been criticized as undermining the political power of ecofeminism. However, King sees ecofeminism's universalist tendencies as unifying: 'politically, ecofeminism opposes the ways that differences can separate women from each other through the oppressions of class privilege, sexuality and race' (1983: 15).

Ecofeminism also tends to essentialize nature itself. It considers nature to encompass all ecological aspects of the environment as well as natural (biological) human needs and capacities. Leach *et al.* argue that 'equating "the environment" with "nature" can obscure the historical and continued shaping of landscapes by people, often within conceptions of society and environment as inseparable' (1995: 3). Such arguments further undermine the overarching view of ecofeminism. It was presented at the 1992 Rio UNCED Global fora as being generalizable to all women. As Braidotti *et al.* (1994: 164) note, essentialist ecofeminism was seen as a source of women's empowerment by reversing 'patriarchal power structures and [placing] women at the top of new gynocentric hierarchies'. However, they concluded that: 'Despite its powerful mobilizing potential, this approach may become a self-defeating strategy, in particular as it has marginalized other approaches in ecofeminism and led to the disenchantment of many women in the environmental movement with associating themselves with ecofeminist positions' (Braidotti *et al.* 1994: 165).

Both Rocheleau *et al.* (1996) and Leach *et al.* (1995) point out the difficulty experienced by many researchers in reconciling ecofeminist views with the everyday situations found in the field. All these authors stress the need to provide a local context for any study of gender and the environment, by contextualizing development in the social and natural environment. Kirk argues that a sense of place is something few ecofeminists address, 'perhaps because many of us live in urban areas or are relatively mobile' (1998: 192). On the other hand, to assume that certain cultural groups have a natural affinity with the land is equally essentialist. Thus ecofeminism suffers from multiple essentialisms, not only of women and of ethnic groups but also of nature/environment itself. An awareness of these underlying assumptions makes it possible to carry out relevant community-based environmental work, while avoiding essentialist arguments about the uniqueness and profundity of land-based local cultures.

Moreover, the association of women and nature is not a transhistorical and transcultural phenomenon. Huey-li Li (1993) points out that a normative link between women and nature is not a cross-cultural belief, since nature as a whole is not identified with women in Chinese society. Yet the lack of the transcendant dualism identified by Western writers does not preclude the oppression of women in Chinese society nor environmental degradation. Furthermore, we are oversimplifying the etiology of environmental problems by blaming men for much that is beyond male hegemony. Postmodern critiques of both ecofeminism and the woman–environment–development debate reject their universalizing tendencies and emphasize the importance of local knowledge and concrete situated experiences in understanding women's connections with the environment.

Feminist political ecology as utilized by Rocheleau *et al.* (1996) brings together much of ecofeminism but takes into account the above-mentioned critiques. It deals with how gender interacts with class, race, ethnicity, national identity and situated knowledge to shape experiences of and interest in the environment. Feminist political ecology, perhaps combined with a materialist ecological feminism, provides a stronger theoretical framework for studies of gender and environment than an uncritical acceptance of the term 'ecofeminism'.

Contextualizing gender differences in environmental awareness

Ecofeminism assumes gendered environmental awareness, yet actual studies of gender differences in concern for the environment have been relatively few, especially of countries in the global South (Momsen 2000). Much of the information that currently exists about such differences is from studies that have examined concerns about *local environmental issues*, which pose a threat to community health and safety (Blocker and Eckberg 1989; Sarin 1995; Shah and Shah 1995). These studies have consistently shown women to be significantly more concerned about such issues than men (Mohai 1992), but Leach *et al.* (1995) argue that such differences are socially constituted. Gender differences in perception of *general environmental issues*, that is problems not specifically limited to those in the neighbourhood or community (Blocker and Eckberg

1989; Momsen 1993) have been less clear and have varied from study to study.

Gender and environmental concern at national and local scales

Gender differences in understanding of the natural environment at different scales were noticeable in a study in Barbados, which focused on soil erosion and was based on interviews with 85 men and 90 women respondents in four communities in the northern part of the island (Momsen 1993). There was little difference in the mean age and education levels of men and women. It was found that 74 per cent of the men, but only 48 per cent of the women, felt that soil erosion was a national problem. At the local scale gender differences in awareness of environmental problems were more marked but the level of such awareness varied. In the most seriously eroded district of Barbados, a lower proportion of men (62 per cent) expressed serious concern over soil erosion as a local problem as compared to the 74 per cent who saw it as a national problem, while there was only a very small difference among the women interviewed (Momsen 1993). Overall, men and women living in the steeply-dissected Scotland District, where soil slippage is sweeping away roads and houses and which has been the focus of large-scale anti-erosion measures for over 30 years, were less concerned about it than those living elsewhere on the island, where soil erosion is more gradual and less catastrophic (ibid.). This unexpected result could be related to the fact that, if your house slid down the hillside the government replaced it and, like the new roads built to replace those washed away, this was generally a big improvement.

Gender roles and environmental concern

The expectation that women are more concerned about environmental problems than men is based on the argument that from childhood on women are socialized to be family nurturers and caregivers, that is to develop a 'motherhood mentality' (Mohai 1992). It has been hypothesized that the attitudes derived from this socialization are reinforced by the roles that women assume in their adult lives as homemakers and mothers. In contrast, in most societies men are socialized to be protectors of and providers for the family.

As in the case of women, the attitudes acquired through socialization may be further reinforced by the roles that men assume in adult life as members of the formal workforce. However, in most societies these traditional gender roles are changing and provide an unstable argument on which to base gender differences in levels of environmental concern.

Whether women in reality are more concerned about the environment than men has not been determined conclusively by empirical studies. Attitudes may be influenced by type of problem as some environmental issues are subtle rather than dramatic. The effects of pollution are often only slowly apparent, with the consequent deterioration in environmental quality more typically showing up in small ways in the ordinary lived environment. As a result of women's social location as managers of the domestic environment they are often the first to notice the effects of pollution. Joni Seager (1996) sees this social role as the main determinant of women's grassroots organizing of environmental protest.

Studies of the effects of parental, homemaker and workplace roles on gender differences in environmental concern have provided mixed evidence. In Barbados (Momsen 1993), I found that there was less difference between male and female farmers in their concern about soil erosion than among the general population. Among non-farming women, all with children, there was more concern about the local environment of the community, especially in relation to pollution from traffic and garbage, than about the general danger of soil erosion to the nation.

Gender differences in knowledge of environmental issues

In the Barbados (Momsen 1993) study and in a Costa Rican study of perception of volcanic hazards (Lemieux 1975), women appeared to be less aware of the causes of problems in the natural environment than men. In Barbados over half of the men but slightly less than a third of women surveyed realized that ashfalls from volcanic eruptions on neighbouring islands were crucial to maintaining the fertility of soils on the island. Women were more likely to say that they did not know the answer to a question, and so men appeared to be more aware of environmental problems than women.

Most studies in the industrialized world have tended to focus on gender roles rather than on the effect of differences in education and decision-making power. In poorer countries these latter issues may be overwhelming. In Barbados, where there are currently more women attending university than men, educational differences are closely linked to particularly age cohorts. We had elderly women telling us that they had no knowledge of a particular issue and directing us to their school-age grandchildren who were very happy to explain the effects of Amazonian deforestation. It may be that women are more likely than men to admit that they do not know the answer to a particular question since lack of such knowledge does not reflect on their status.

Other studies of gendered links between information and environmental concern also question the extent to which they are meaningful. In hazard perception, vulnerability is an important distinction and women are often the most vulnerable, since they are less likely to have the resources to recover from a hazardous event or the community status to be able to obtain assistance. This is clearly seen in a study of hazard perception in an area affected by volcanic activity in Costa Rica (ibid.). Men were more likely to feel that they could depend on official assistance but they also stated that they believed (falsely) in the efficacy of specific predictors of future eruptions. As in both the studies of soil erosion and water conservation (Table 5.1) in Barbados, women were more likely to say that they did not know, thus giving the impression that men were more environmentally aware. One Costa Rican woman in her forties stated, in response to a question, that she had had 15 children and could no longer think!

Table 5.1 *Gender and the meaning of water conservation in Barbados, West Indies*

Meaning	Male		Female		Total	
	No.	%	No.	%	No.	%
Water conservation is:						
saving water	11	40.7	16	59.3	27	27.3
reducing wastage	11	55.0	9	45.0	20	20.2
limiting use	8	53.3	7	46.7	15	15.2
optimizing use	6	66.7	3	33.3	9	9.1
do not know	2	8.3	22	91.7	24	24.2
other	4	100.0	0	0.0	4	4.0
Total	42	42.4	57	57.6	99	100.0

Source: Griffith (2001).

Among West Indian small farmers the ranking of environmental concerns reflects gender differences in reproductive roles as well as environmental differences between islands. Women farmers undertake most farm tasks but always seek male labourers to spray pesticides and herbicides for them, as women have long been aware of the dangers of the use of agricultural chemicals, especially to pregnant and nursing mothers (Momsen 1988b; Harry 1993). In Trinidad, as early as 1978, women farmers were also expressing a fear of chemical contamination of food and were selling organic produce in village markets at a premium price (Harry 1980).

Structural and situational gender differences

The influence of state policy is clearly seen in our work in Yunnan, China, where the national government had been pursuing a publicity campaign stressing the damage that deforestation could do to the environment. In the four mountain villages surveyed, this message was reinforced by severe punishment for transgressors. It was most accepted in those villages where it was reinforced by the village headman's commitment to protect the local environment and natural resources. In the most isolated village, where the official forest guards were least likely to be present, there was a great deal of illegal logging for sale, though it was often not done by the villagers themselves. Men and women in Yunnan seemed to have equally absorbed the government message. The only informant who felt that logging should be allowed was a woman whose husband was in jail for such illegal activity (Momsen 2000).

In all societies gender roles are changing. In industrialized countries it is getting less meaningful to separate homemaker and paid-worker roles. In Barbados and China, fieldwork revealed a decline in the specificity of gender roles, with more household and farm tasks becoming gender neutral. As roles change and become less gender-specific many of the materialist arguments for women's greater awareness of environmental problems at the local level become less persuasive.

Gendered use of natural resources

Water is essential to human life and in most societies women are responsible for supplying it to their families for drinking, cooking,

Box 5.1

Water and sustainable development

Over one billion people, or 18 per cent of the world's population, lack access to safe drinking water and over 2.4 billion people are without basic sanitation (Lefèbvre 2002a). The target of reducing by half by 2015 the number of people without access to clean water and sanitation was adopted in the final Plan of Action at the World Summit on Sustainable Development, held in Johannesburg in 2002. According to the United Nations Environment Programme (UNEP), every year over two million people, mainly children, die from diarrhoea due to contaminated water and lack of sanitation, and five million deaths are caused by water-borne diseases (ibid.). Water shortages are increasing and 30 countries, mainly in Africa and the Middle East, are facing water deficits already (ibid.). The United Nations believes that at the current rate of consumption two out of three people will be living in water-stressed situations by 2025. There is plenty of water available on the Earth but it is not evenly distributed. Consumption also varies widely: on average, Americans use 700 litres of water per person per day, Europeans 200, Palestinians 70 and Haitians less than 20 (ibid.). Providing water for the family is usually women's task, with help from children, and so improved access will reduce their burden of work, freeing time for other occupations and more regular attendance at school, as well as improving health.

Aid agencies agreed in Johannesburg to improve coordination, utilize partnerships and respond to requests from beneficiary countries for water projects. Hopefully this new approach will reduce the problems experienced in implementing such projects. Currently, aid agencies usually provide boreholes and offer protected wells or handpumps to overcome the problem of water contamination. Theoretically, a pump can service an entire village for 14 years, but because of shoddy installation a pump may fail within days. Wear and tear on pumps are inevitable, but the donors like to move on to new projects and leave repairs to the villagers, but many communities are too poor and disorganized to do this. They may be too isolated to be able to get the spare parts and may not have the knowledge to install them. Wells are easier to maintain but donors and local authorities tend to prefer the more sophisticated higher technology of pumps because they feel it is progressive and modern. Protected wells are more vulnerable to contamination than handpumps but cost only a quarter of the price of a pump, are more sustainable and can dramatically reduce the incidence of cholera and diarrhoea.

In studies in Ethiopia, Ghana, India and Tanzania it was found that time previously spent on water collection was diverted to income generation, observance of social obligations and attendance at school. Women and girls benefited most. Academic performance improved because children stayed longer in class and teachers did not have to fetch water for them. Death and disease has been reduced and in some cases gender roles have become more interchangeable as women have become empowered.

Source: adapted from Lefèbvre (2002a); Carroll (2002b); Pickford (2000).

cleaning, bathing and clothes washing. It is also needed for watering animals and for irrigating crops. Water is becoming a scarce and increasingly polluted commodity in many places as population density and use of agricultural chemicals increases (Box 5.1). Over a billion people lack a safe water supply and in 2002 the United Nations added the unfettered access to clean water to its international covenant of economic, social and cultural rights.

As the daily search for pure drinking water becomes a more time-consuming task in many parts of the world, less time is available for other household tasks, such as childcare and cooking (Box 5.1). Availability of water varies with the seasons: in the dry season in many places rivers and springs may dry up and the search for water becomes even more difficult and contamination more likely. In addition, irrigation schemes, where there is standing water in ponds, may increase the prevalence of mosquito-borne diseases, such as malaria, yellow fever and dengue fever. In many cases the needs of women for access to water are ignored when irrigation schemes are planned, and women are rarely included in the community irrigation management group. Women may not attend meetings concerning irrigation schemes because they do not have time, cannot leave children or are not allowed to appear in public. Planners often assume that male heads of household will represent the needs of all members of the household. On the rare occasions when women are officially involved in irrigation projects, they may be too shy to speak out and/or may be ignored when they try to speak in public.

The use of polluted sources of water results in the spread of water-borne diseases such as cholera, typhoid and amoebic dysentery, as well as stomach upsets and diarrhoea. Development interventions by supplying wells to villages have reduced the spread of these diseases but have brought their own problems (Box 5.1). The introduction of charges for use of the new well may drive poor families to return to using their old polluted source, while in Bangladesh the new tubewells were found to be contaminated by arsenic from the underlying geology of deltaic deposits. Thus in West Bengal, India, and in Bangladesh slow arsenic poisoning has become a major environmental disaster. Good nutrition provides some protection from this poisoning but most villagers, especially women, in the region are undernourished. Nor can they afford to buy filters which may remove the arsenic from the water (Box 5.2).

Box 5.2

Arsenic poisoning in Bangladesh

When her mother died in May 1999, at the age of 26, after months of sickness from drinking water containing high levels of arsenic, Shapla, aged 10, who already showed signs of the skin lesions and sores of arsenic poisoning, worried about the survival of her malnourished seven-year-old brother and her nine-month-old sister. By June 2000 her sister was dead. Her father, a rice mill worker, was too poor to purchase filters to remove the arsenic from the village water and vitamin tablets to protect his family. His wife had been recovering from arsenic poisoning but succumbed after the local doctor refused to continue treating her without payment. His mother, aged 55, was also suffering from arsenic poisoning and was baffled by this illness which had attacked 10 of her 13-member family. A number of arsenic-contaminated tube wells in the village had been sealed, without providing alternative sources of water, thus forcing women to expend more energy and travel further in search of clean drinking water.

Bangladesh is currently in the middle of what the World Health Organization calls 'the largest mass poisoning of a population in history'. Some 40 million people, most of whom live in poor rural areas, are exposed to arsenic-contaminated water. At least 59 of the nation's 64 districts, in which 80 million people live, have arsenic-contaminated groundwater. Possibly 30 per cent of tubewells, the provision of which was completed in the early 1990s, have unacceptably high levels of arsenic. People are affected in different ways and some appear to be relatively resistant. It can take from 2 to 15 years for symptoms to appear, but gradually victims, mainly in villages, become lethargic, weak and unable to work. In this way, arsenic poisoning is a threat to food security in Bangladesh and may soon become an epidemic. In addition there is fear that arsenic may be getting into the food chain through contamination of irrigation water.

Many Bangladeshis find it hard to understand that water that looks clean and is tasteless can cause such problems. It has been called the worst environmental risk ever, worse than Bhopal or Chernobyl. Predictions of deaths from causes related to arsenic are difficult, but vary between one and five million people. Rural people, because of ignorance, consider arsenicosis a curse of nature and fear that it is contagious. The social impact is enormous, with girls showing signs of arsenic poisoning being unable to find husbands and married women being rejected by their spouses. Families often force victims to live in isolation or remain within the house and so patients complain of loneliness. People from villages with high levels of contamination may be shunned as marriage partners and as employees. A study in three badly affected villages found that prevalence rates for women were higher than for men and that more women (64 per cent) than men (36 per cent) thought it was contagious (Bhuiyan 2000). Higher rates of arsenicosis in women are related to their lower nutritional and educational levels. Among the 291 cases of arsenic poisoning in the three villages, two-thirds complained of social and psychological problems, of whom 64 per cent were women (ibid.).

Source: compiled from: Bhuiyan (2000); Chowdhury (2001: 67–89); Ahmed, (2000); Bearak (2002).

In many places water is generally becoming scarcer and conservation is vital. Barbados, a Caribbean coral island, draws its water from underground sources. The water is derived from rainfall and is pure because it is stored as subsurface groundwater and is filtered by percolating through the coral. However, the local demand for water is increasing as most people now have piped water in their houses and access to flush toilets. There is also high demand from tourists, who tend to shower more frequently than local people, expect to have their sheets and towels changed daily, and want to see the grounds of their hotels and their golf courses green and lush at all times of year. However, unless there is a clear understanding of the need for water conservation by both locals and visitors it will be impossible to reduce demand to any great extent. A recent study has shown clear gender differences in this understanding, with women who are likely to be the heaviest users of water showing the least knowledge (Table 5.1).

The availability of pure drinking water may also be a problem in non-tropical countries, especially in post-communist countries, where pollution and decaying infrastructure are widespread. In the Ukraine, MOMA-86, a women's NGO founded as a response to the Chernobyl nuclear disaster of 1986, is the foremost environmental organization in the country (Khosla 2002b). In 1997 MOMA-86 initiated a drinking water campaign. This campaign aimed to find solutions to drinking water problems at the local level through water quality monitoring, raising public awareness on water and health problems, sustainable water management and environmental rights and promoting public participation in planning. In towns, the main drinking water problems are the low quality of the water and water shortages, because most of the water delivery systems are worn out and leaking. In the countryside wells have become contaminated by nitrates and other chemicals. MOMA-86 estimates that 45 per cent of the population is consuming water that does not comply with government standards (ibid.). At the same time tariffs for water have risen rapidly and pensioners may be paying up to 15 per cent of their incomes for water (ibid.). There has been an increase in diseases caused by polluted water and these are particularly affecting children. MOMA-86 works actively with women as they feel that women are the most concerned group, particularly where children are becoming sick. MOMA-86 is also working with community groups to find alternatives to polluted sources, has installed water purification devices in kindergartens, has lobbied to get a new law passed on drinking water supply and is running education campaigns to encourage people to save water.

Until recently, water resources were considered unlimited and obtainable for free or for only a nominal sum. Growing awareness of water pollution and frequent absolute shortages have created new pressures to protect water supplies. Women are often most immediately affected by these changes because of their household responsibilities involving water and their role as caregivers for the sick in the family (see Plate 5.1).

Forests and woodlands

Forests are sources of timber for construction and manufacturing and for fuelwood. Leaves provide feed for stock, bedding and organic material for gardens. Forests are also important as sources of wild plants which provide supplementary nutrition, such as trace elements and vitamins, otherwise missing in the diet. Wild plants may also have medicinal uses, while some forest products, such as mushrooms, honey or chicle, may be sold for extra income. Forested areas may also be sources of meat from animals like wild boar or deer.

Plate 5.1 *India: women using a village water pump.*
Source: Janet Townsend, University of Durham, UK

Population growth and demand for tropical hardwoods from the North have led to the rapid depletion of many forests. In many areas growing commercial use of timber has led to the privatization of forests. Generally, men are involved in timber extraction for construction, charcoal-making and the sale of firewood to cities. Women's use of forests tends to involve more subsistence demand. Women collect wood for fuel and forest plants for use in the home but may also sell some forest products. The decline of communal forest lands has forced women to walk further in search of fuelwood and to make do with types of wood that do not burn well or that produce a lot of smoke (Box 5.3).

In a study undertaken in Ghana in 1983 (Ardayfio-Schandorf 1993a) it was found that, in comparison with 1973, fuelwood was more scarce, especially good burning species, women had to walk further to find it and, if they had to buy it, as they did in coastal villages, it was more expensive (Figure 5.1). In Sri Lanka in 1950 women obtained all their fuelwood from the forest, but by 1988 this resource

Box 5.3

Kenya: women's role in reafforestation

In 1989 Professor Wangari Maathai was awarded Woman-Aid's 'Woman of the World' award. This was in recognition of her work in setting up the Green Belt Movement (GBM) in Kenya in 1977. Working with the National Council of Women she persuaded communities throughout Kenya to plant more than 10 million trees. Some 35 other African countries have taken up the scheme. Thousands of green belts have been planted and many hundreds of community tree nurseries set up. Women have shown each other how to collect the seeds of nearby indigenous trees, and how to plant and tend them. Slowly the devastating effects of soil erosion are being reversed. The GBM is a grass-roots environmental movement with multiple objectives: to reduce deforestation by planting trees; to promote the cultivation of multipurpose trees; to prevent the extinction of indigenous species; to increase public awareness of environmental issues; to create a positive image of women; to make tree-planting an income-earning activity for women and to help the rural poor.

Maathai started the movement, despite opposition, because she realized that much of Kenya had already been cleared of trees and bushes and that more cash crops would only accelerate the process of desertification. There was little firewood left to gather and rural Kenyans were forced to depend on agricultural residues and dung for cooking and heating and so had to eat an increasing number of highly processed foods. In 2003 Professor Maathai's many years of environmental leadership were recognized in her own country by her appointment as deputy minister of the environment by President Kibaki.

In addition to the Green Belt Movement, local authorities, village chiefs, schools and prisons have established seedling nurseries in Kenya. Women are motivated to grow trees not only for fuel but also for fodder and fruit, for use as windbreaks, for fencing and for shade and construction materials. In a study of women's community forestry practices in part of central Kenya (Hyma and Nyamwange 1993), it was found that women had many reasons for participating in tree planting: they recognized the benefits of trees to soil and water conservation; they saw the utility of trees in generating income and providing for household needs in terms of fibre, fuelwood, shelter and medicine; and they wanted to preserve indigenous species. Women have been encouraged by the public recognition of their activities, by an increase in extension workers and services tailored specifically to women's needs and by the free supply of tree seedlings. A major role of the GBM has been the provision of technical assistance at national and international levels to other community groups.

Hyma and Nyamwange (1993) also found that many women's forestry groups suffered from disorganization. Heterogeneity both within and between groups, with respect to age, education, status and motivation, also caused problems. Constraints on tree-planting identified by women's groups included the following: lack of inputs such as seedlings, containers and fencing; plant infestations; shortage of water and manure for seedlings; and lack of training, management skills, time, land and capital. There are very few trained women foresters and it was felt that tree nurseries needed to have paid staff rather than depend on women's volunteer work. Planning programmes need to reflect the existing indigenous knowledge of tree management, conservation practices and interest in and needs for different trees of women and men, rather than seeing tree planting as yet one more responsibility of rural women.

Sources: based in part on Vidal (1989) and Hyma and Nyamwange (1993).

---

yielded no fuelwood and the new sources were common lands (58 per cent), own home gardens (40 per cent) and farmland (2 per cent) (Wickramasinghe 1994). In Bangladesh the percentage of women using different fuels was as follows: dried cowdung (60), agricultural residues (48), twigs and leaves (47), waterplants and biosoil (41) and wood (35) (Rumi and Ohiduzzaman 2000). Substitution of high-energy fuelwood with low-energy species or other types of fuel increased the workload of women (Ulluwishewa 1993). Chopping low-energy fuelwood into manageable pieces, to allow it to dry and so reduce the moisture content, demands additional inputs of time and energy from women. Unless the moisture is removed, lighting low-energy fuelwood is difficult. In the wet season it is hard to collect and to dry wood, agricultural residues or cowdung. Such seasonal shortages can be overcome by purchasing fuel if the family can afford it, but landless families in Bangladesh had an average of

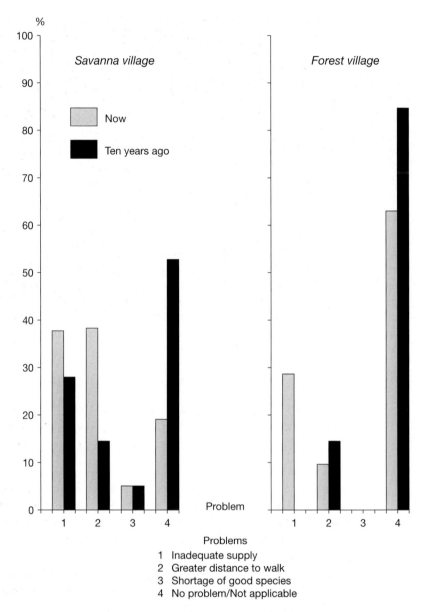

Figure 5.1 *Ghana: problems of fuelwood collection in different ecosystems.*
Source: Ardayfio-Schandorf (1993)

156 days a year without fuel for cooking (Rumi and Ohiduzzaman 2000). It was also found that, as women had to walk further to collect wood, they reduced the number of trips per week from five to two, but increased the weight per load from an average of 8.3 kg to

Plate 5.2 *Ghana: a group of women walking along a road carrying wood for fuel on their heads.*

Source: author

27.6 kg. In Sri Lanka the distance walked to collect fuelwood increased from an average of 0.25 km to 3.3 km, and the time taken searching for suitable wood also increased from 30 minutes to 4.5 hours (Ulluwishewa 1993, 1997).

According to the United Nations Development Programme (UNDP), women and young girls in sub-Saharan Africa carry, on average, more than 20 kg of wood over 5 km every day (Lefèbvre 2002b). Where women have to seek firewood beyond the area that they perceive to be safe they need to be accompanied (Plate 5.2). Sometimes, if the family owns a bicycle, husbands will help in transporting the fuelwood as women do not ride bicycles (see Plate 5.3). In Bangladesh women face social censure if they venture away from the home to collect firewood. This problem is most severe for poor women from landless families, while large farmers do not have this problem as they do not need to venture off their own land (Rumi and Ohiduzzaman 2000). Women also expect children to help, so that in rural Bangladesh 10 per cent of children are forced to miss school, but the supply of fuelwood remains the woman's responsibility (ibid.).

Plate 5.3 *Brazil: a man collecting mangrove wood and transporting it on a bicycle in Maranhão.*

Source: author

The use of low quality fuel for cooking means spending more hours by the side of the fire, usually in a poorly ventilated kitchen, where inhaling acrid smoke is hard to avoid (see Plate 4.2). It has been calculated that a day spent inhaling smoke from cooking fires is the equivalent of smoking 400 cigarettes and can cause chronic respiratory problems, throat cancer and stillbirths. Women who cooked with straw or wood when they were 30 years old were found to have an 80 per cent greater chance than other women of having lung cancer in later years (Ardayfio-Schandorf 1993). The World Health Organization (WHO) estimates that 2.5 million women and children in developing countries die prematurely from inhaling toxic fumes from the biomass fires used indoors to cook their food (Lefèbvre 2002b).

However, smoke is seen by many women as an essential element for food storage as it reduces the losses to rats, insects and fungi. Grains, pulses, seeds and various kinds of wild fruits, meat and fish are stored above the fireplace, where they are kept warm and dry. In thatched-roof huts the smoke can escape through the thatch and also repel insects and pests. Smoke becomes more of a health hazard when thatched roofs are replaced by tile or corrugated iron.

As the cost and effort of obtaining fuelwood increase women are compelled to economize on its use. They adopt a range of strategies to achieve this:

- Women move from cooking outdoors to indoors to minimize the loss of heat from wind.
- Most women are aware that improved stoves, which are enclosed and have an opening on one side for insertion of wood and holes for the pots on top, are more efficient and economical than the traditional three-stone open fire (see Plates 4.2 and 6.7). Such stoves have been widely adopted in Asia and Africa. In Kenya the 'Jiko' stove, consisting of a tin can with a ceramic lining, was introduced and is now used by half the poor families in the country. This costs between US$2 and US$5 and saves 590 kg of fuel a year, worth about US$65, a big saving in time and money for women (Hesperian Foundation 2001).
- When fuelwood is abundant the fire is kept burning to provide protection from mosquitoes and wild animals, but with scarcity fires are extinguished immediately after cooking.
- Aluminium cooking pots are energy efficient so they are gradually replacing earthenware pots in order to reduce the use of fuelwood. But aluminium pots are seen as being highly priced, earthenware vessels keep food warm for longer and many people say that the food cooked in the traditional pots tastes better (see Plate 5.4).
- When there is no shortage of fuelwood, water is heated to bathe children, the old and the sick. Drinking water is also boiled to purify it. With fuel scarcity cold water only is used for such purposes. Drinking unboiled water is likely to lead to sickness, especially in young children, pregnant women and the elderly, leading to greater stress in the family. Washing clothes and dishes and bathing babies in cold water is harder work for women.
- At busy times, women cook more food than is required for a single meal. The leftovers are used for a second meal. With plentiful wood the leftovers would be warmed up, but when wood is scarce and expensive they are served cold, saving both time and fuel. In tropical climates without refrigeration left-over food quickly becomes contaminated. Women usually eat last in Asian and many African families, after serving other family

Plate 5.4 *Ghana: a woman cooking over an open fire in a compound in the north. Note the enamel bowls.*

Source: author

members, and so are likely to get the smallest portions and also the food that has been left out longest.

- Carrying heavier loads of wood longer distances is a hardship for women.
- The extra time needed to find wood and the lack of fuel for cooking forces families to reduce the number of meals they consume each day in the most acute cases.
- Scarcity of fuel forces families to cut down on consumption of food items needing long cooking times, such as pulses and yams, and to give up smoking some items for longer preservation (Bortei-Doku Aryeetey 2002).
- Wood ash from fires is traditionally used to fertilize dooryard gardens but the supply of this decreases with declining use of fire.
- There has been a shift from the use of traditional herbal medicines to Western medicines, in part because of the time and amount of wood needed to prepare herbal remedies.

- In many parts of Africa women are the main producers of local wine or beer. Such products are often the only source of cash income for women. However, brewing of alcoholic beverages takes a lot of fuel and so women may be forced to cut back production (see Plate 6.7).

In response to the fuelwood crisis, international agencies developed programmes of social or community forestry in many parts of the South. These were specifically aimed at helping the rural poor and later developed a focus on poor women as the main beneficiaries. On the whole, these projects failed to help women because they were 'top-down', involving paternalistic attitudes to the poor, overly centralized planning, poor delivery of support services, elitist attitudes, especially among poorly trained government foresters, and unsuitable technology such as concentration on ecologically unsuitable exotic species like eucalyptus (Gain 1998). Community resource management institutions in India, formed as part of the Joint Forest Management (JFM) programme in the 1980s, were celebrated as a success but Agarwal saw them as 'gender exclusionary and highly inequitable' (1997a: 1374).

In many cases these projects tend to benefit the richest families in the community and women's needs are not always taken into consideration. However, women are often employed in tree nurseries as this is considered suitable for women because of their traditional nurturing roles. Such tasks may be seen as providing additional income to women, or merely as consuming more of women's scarce time for something that does not benefit them in the long run, if the trees are planted on men's land and sold rather than used for firewood (Bortei-Doku Aryeetey 2002)

One quarter of the world's population does not have access to electricity and the annual expansion of new connections does not keep up with population growth. Some 2.4 billion people depend on biomass for heating and cooking and the number is increasing (Lefèbvre 2002b). The World Summit on Sustainable Development in 2002 agreed that meeting the goal of halving poverty levels by 2015 could not be achieved without improving access to energy supplies. It was also noted that the impact of energy scarcity on women was a major contributor to the problem of gendered inequalities of opportunity (ibid.).

Gendered impacts of natural hazards

Vulnerability in combination with the occurrence of a natural hazard produces a disaster (Wisner 1993). Poverty is one of the main aspects of vulnerability but it varies with occupation and social characteristics, such as gender, age, ethnicity and disability. The gender impact of natural hazards, such as volcanic eruptions, earthquakes, hurricanes, typhoons and floods, reflects the position of women in different cultures. Women generally have less access to resources and less representation at all levels of decision-making. Women may suffer more than men in most disasters but may also have central roles in coping and recovery, providing that they are given the opportunity (Jiggins 1986; Rivers 1987; Rashid 2000). Women may be forced to turn to high-risk activities like prostitution in order to feed their children following a disaster, while men can migrate alone in search of employment. Women on their own are even more vulnerable and may find it especially difficult to get loans for rebuilding and re-establishing a viable livelihood (Wisner 1993).

In the Costa Rican study mentioned above, women were less likely to get state aid as they had fewer contacts in positions of power who could help them and were less used to dealing with outsiders and seeking help. Thus their main recourse in the face of disaster of praying to God was rational given their powerless position in terms of human assistance. In the predominantly Muslim society of Bangladesh many women are not allowed to speak to strange men or be in public without a male relative. It is a women's duty to protect her home and her children (see Plate 5.5). Thus when there are severe floods, which occur regularly in this deltaic nation, women will not leave their homes to go to shelters in case their husbands accuse them of not looking after their homes. At the same time they are afraid of the relatively public space of the shelters and especially fear for the safety of their daughters there. So sad incidents occur, such as that of the woman who scrambled into the roof space of her home clutching her baby as the water rose beneath her and fell asleep, her baby slipped from her grasp falling into the water below and drowning (Hussain 2002). During floods the normal gender division of labour is not changed and women's tasks, such as carrying water, cooking and caring for children and animals, become very difficult. In addition, women's assets, such as milch cows, cooking utensils and poultry, are very vulnerable. Furthermore,

Plate 5.5 *Bangladesh: a woman applying a new layer of mud to strengthen the foundation of her house during the monsoon season, when flooding is common. Her husband is a handloom weaver as the family has no land, so the wife is responsible for all outdoor work as well as housework and childcare.*

Source: Michael Appel, Davis, California

while moving about during floods 'the sari is like a death trap for women' (Nasreen 2000: 314–15).

When assistance is sent to flooded areas, women are often last to get it as the men push them out of the way in the rush to grab supplies. Also women may have lost clothing in the flood, are thus unable to cover themselves adequately and so cannot enter public areas. Pregnant women may miscarry or give birth prematurely or have other medical needs for themselves or their children but will not consult male doctors. Clearly, if women are to receive adequate disaster assistance there need to be more women doctors and volunteers and more women-specific aid provision, such as clothing, breast pumps for nursing mothers who have lost their babies, diapers (nappies) and adequate safe and private sanitation facilities for women. Poor women know fewer languages, so may not be able to communicate with aid workers and, when given unfamiliar food in aid packages, may not know how to cook it. However, gender discrimination can make it difficult for women relief workers to do their jobs and to change the way aid is allocated (Begum 1993). In environmental disasters women in Bangladesh are made more vulnerable by their social status and gender role. In the 1991 flood it was estimated that 85 per cent of the deaths were of women and children. Famine and epidemics often follow floods and poor women cope by utilizing aquatic plants and also flowers and stems as well as fruit from homestead trees. Women were found to be most knowledgeable about such famine foods (Hossain 2000).

In the Caribbean island of Montserrat, when a hurricane destroyed much property, there was a shortage of construction workers so young women learned how to be tilers and plasterers. In this way, short-term demand enabled a change in the traditional division of labour and women gained a new skill which was of long-term utility. After Hurricane Mitch in Nicaragua in 1998, aid workers tried to ensure that both women and men got access to assistance. By linking this assistance to gender workshops they hoped to improve mutual understanding within families and communities, ease trauma and, by helping men to appreciate the work done by women, reduce the extent of domestic violence. Despite this emphasis on gender awareness, it was young women and men who appeared to have benefited most from reconstruction projects and the impact of 'masculinity' programmes on gender relations was unclear (Bradshaw 2001).

Environmental problems in urban areas

Cities in the South are growing rapidly as people leave rural areas in search of better paid jobs and the bright lights and opportunities of the city. In rapidly urbanizing developing countries the rate of urban growth is so fast that city services cannot keep up and the consequent deteriorating urban environment results in increased disease and ill health. Most migrants end up living in slums with no paved roads, running water, proper sewage or solid waste disposal facilities. These shanty towns are often built on unsuitable land, blocking natural drainage channels and so increasing flood danger, as in Dhaka, the capital of Bangladesh, where rubbish dumps and the recently banned plastic bags also block drainage. On steep slopes in Rio de Janeiro, Brazil, where the *favelas* that are home to the poor are built, heavy rains often cause landslips, bringing houses down with them. Tapping illegally into high voltage electricity wires by slum dwellers can lead to death from electrocution and power outages. Such damage to wires led to authorities in Mexico City providing free access to electricity in poor areas, as this was cheaper than constantly having to repair the overhead wires. Overcrowded housing conditions, lack of sanitation and poor drainage, plus heavy traffic and inefficient industries, add to the environmental problems of air and water pollution. Dealing with such conditions at the household level is usually the responsibility of women, and the constant dirt and presence of rats, cockroaches, flies and mosquitoes make housework and childcare especially onerous (Simard and De Koninck 2001). A study of the capital city of Ghana, Accra, revealed that environmental health problems were most acute in the rural fringe squatter settlements and other residential areas occupied by the poor, particularly in terms of childhood diarrhoea (Songsore 1999). Cooking on both wood and charcoal exposes users to high levels of particulates and carbon monoxide. Only the rich have the luxury of separate kitchens and the use of electricity and gas, so it is not surprising that children and women in poor households demonstrate high proportions of respiratory illness when compared to middle- and upper-class areas in Accra (ibid.).

Despite strong theoretical arguments suggesting that women are more protective of the environment and more aware of environmental problems, investigations into gender differences in concern for the environment have been relatively few and are generally inconclusive in their findings. Where differences do exist, they do not show a consistent pro-environment stance for either gender. Furthermore, statements about such concern may reflect gendered social roles and positions more than real differences. It may well be that contemporary economic development pressures on natural resources, changing gender roles and wider access to education are undermining the long-established patterns of gender differences in environmental awareness that formed the basis of ecofeminism. It appears that a strongly place-based feminist political ecology approach may provide a better understanding of gendered environmental perception than ecofeminism.

Learning outcomes

- Ecofeminism has a plurality of positions: cultural, liberal, social and socialist.
- Alternatives to ecofeminism include feminist political ecology, feminist environmentalism and development policies based on gender and environment rather than women and environment.
- The use of natural resources can be gendered and the privatization of common resources impacts women's access.
- Women are more vulnerable to natural hazards due to cultural restrictions and social status.

Discussion questions

1 Explain how ecofeminism can be considered essentialist.

2 Why do women sometimes cook on wood that produces a lot of smoke or does not burn well?

3 Why do women more than men admit ignorance of causes of environmental problems?

4 Discuss the gendered impact of water pollution in poor countries.

Further reading

Quarrie, Joyce (1992) *Earth Summit 1992*, London: The Regency Press Corporation. Provides an accessible source of documents from the Rio Conference.

Rocheleau, Dianne, Barbara Thomas-Slayter and Ester Wangari (eds) (1996) *Feminist Political Ecology: Global Issues and Local Experiences*, London and New York: Routledge. An edited collection of case studies of women's activism for protection of the environment. The editorial introduction provides an analysis of feminist political feminism.

Sachs, Carolyn E. (ed.) (1997) *Women Working in the Environment*, London and Washington, DC: Taylor & Francis. Includes case studies of gendered use of natural resources, mostly in developing countries. The editorial introduction provides a review of gender and the environment. Usefully read in conjunction with the book by Rocheleau *et al*.

Shiva, Vandana (1989) *Staying Alive: Women, Ecology and Development*, London: Zed Books. The classic presentation of ecofeminism. Shiva's work led the way in considering women's connections to the environment in the South.

Websites

www.biodiv.org   Convention on Biological Diversity, full text.

www.igc.apc.org/ea/susdev/agenda 21.htm   Rio Agenda 21.

www.igc.org/habitat/agenda21/rio-dec.htm   Rio Declaration, with 27 principles defining the rights and responsibilities of nations in enabling a new and equitable global partnership for development.

www.johannesburgsummit.org   Official documents of the World Summit on Sustainable Development.

www.nt1.ids.ac.uk/eldis/gender/gen_lele.htm   Eldis is a means of accessing on-line information on development and the environment and focuses on countries of the South.

www.unfccc.int   Framework Convention on Climate Change (UNFCCC).

www.wedo.org   Women's Development and Environment Organization (WEDO), an international advocacy network that seeks to increase the power of women worldwide as policymakers in governance and in policymaking institutions, forums and processes at all levels, to achieve economic and social justice, a peaceful and healthy planet and human rights for all. Organizes Women's Caucuses at UN conferences and other intergovernmental forums to coordinate political action.

www.weimag.com   *Women and Environments International*, a magazine produced in Canada and founded in 1976, examines women's multiple relationships to their environments – natural, built and social – from feminist perspectives.

Gender in rural areas

Learning objectives

When you have finished reading this chapter, you should be aware of:

- **impacts of development on gender roles in agriculture**
- **regional differences in gendered employment in agriculture**
- **time use in rural areas**
- **new types of rural employment.**

One of the major demographic trends in the countries of the South is the movement of people from the countryside to the cities (see *Population and Development in the Third World* by Allan and Anne Findlay in this series) but this movement is sex-specific and the predominant group varies from country to country. In Latin America more women than men move to cities, leaving a masculine sex ratio in the rural population, with 106 women per 100 men in urban areas and 90 women per 100 men in rural areas (United Nations 2000). In South Asia and China the nationwide masculine sex ratios are most marked in the rural areas. In Africa and in the Middle East migration to cities is predominantly male and a higher proportion of women stay in the countryside. Rural to rural population movement also occurs. Where this is to a frontier of settlement as in Amazonia (Plate 2.1), then families generally move as a unit, but eventually women and children may be left alone on their holding while men seek paid employment elsewhere. In the countries of Central and Eastern Europe counterurbanization is beginning, as people leave the cities to live more cheaply in rural areas by growing their own food and using wood for heating instead of expensive electricity. Population in sub-Saharan Africa and in South, South-east and East Asia is still predominantly rural.

Twice as many women as men work in an agriculture-related activity in developing countries (Odame *et al*. 2002). In 2000 it was estimated that there were almost six million women directly employed in agriculture worldwide. The numerical importance of women in the agricultural workforce is increasing in developing countries, where the proportion of women workers has steadily increased from 38.59 per cent in 1950 to 43.83 per cent in 2000 and is predicted to rise to 44.44 per cent by 2010, whereas in developed countries the proportion of female agricultural workers has declined and was 36.13 per cent in 2000 compared to 43.77 per cent in 1970 (FAOSTAT 2002). Women's work in agriculture is largely unremunerated and is so undervalued and often unrecorded. It involves not only working in fields and caring for livestock but also post-harvest processing (Plates 6.1 and 6.2), storage of crops and animal products (Plate 8.4), seed selection (Plate 6.8) and marketing (Plate 1.2).

There is gender bias in terms of ownership of resources such as land, and in access to training and modern inputs. In Latin America and the Caribbean in 1993, the proportion of women farmers receiving technical assistance from agricultural extension workers was less than 10 per cent and in most countries less than two per cent (Kleysen and Campillo 1996). Various studies have shown that, if women farmers had access to land of equal fertility to that owned by men, and were able to utilize the same information and inputs and had more control over their own time, then yields from women's farms would be equal to or greater than those from men's farms. Women are most likely to be agriculturists in the poorest countries but there are very distinct variations at the continental scale. The female participation rate in the agricultural labour force is highest in sub-Saharan Africa, Asia and the Caribbean, and lowest in Latin America (FAO 1993) (Plates 6.3 and 6.4).

At the end of the twentieth century only a few countries had time series data on the gendered pattern of employment in agriculture (World Bank 2001). The proportion of the labour force employed in agriculture fell between 1980 and the late 1990s throughout the world, except in some of the countries in transition, where it

Plate 6.1 *Bangladesh: a woman stripping jute fibre from its outer covering.*

Source: Rebecca Torres, University of East Carolina

Plate 6.2 *China: women drying and winnowing rice in Yunnan in the south-west.*

Source: author

Plate 6.3 *India: women planting and a man ploughing in paddy fields in Mysore.*

Source: Janet Townsend, University of Durham, UK

Plate 6.4 *Ghana: a group of women digging with hoes.*

Source: author

increased for both men and women. In Latvia such an increase was encouraged by an early break-up of the Soviet-era collective farms, the widespread reallocation of land and a national history associating peasant farming with independence. In Romania and the Kyrgyz Republic, the proportion of both men and women employed in agriculture and forestry increased during the 1990s as many people were forced to turn to subsistence production for survival. There were marked declines in the employment of women in the agricultural labour force in Jamaica from 23 to 10 per cent, in Malaysia from 44 to 15 per cent, in Honduras from 44 to 8 per cent and in Peru from 25 to 3 per cent (ibid.). In these countries women moved out of agriculture into the service sector (tourism and data entry in Jamaica or manufacturing in foreign-owned factories in Malaysia and Latin America) as new opportunities for non-agricultural employment became available. In some cases this new employment was located in rural areas but in many cases it involved female migration to cities.

There is little recent data for African countries but it is probable that the proportion of women working in agriculture has not changed much. In most countries more men than women work in agriculture, forestry or fishing but this is not true in Africa, where agriculture occupies a much higher proportion of the female labour force than of the male labour force, as it also does in Turkey, South Korea, Vietnam and Romania. Generally, in developing countries, women do about 10 per cent more work than men in rural areas but only slightly more than men in urban areas (ibid.).

Statistical evidence on gender roles in agriculture is very unreliable. In many societies it is culturally unacceptable both for a woman to say that she does agricultural work and for the census taker to consider that she might have an economic role. Detailed fieldwork has often indicated a much higher level of female participation in agriculture than is generally recorded in censuses. In Latin America official estimates of the proportion of women in the rural labour force in 1994, as compared to other studies, were much lower: in Colombia the official figure was 26 per cent as compared to the studies' estimate of 51 per cent; in Peru the figures were 45 and 70 per cent respectively; and in Costa Rica 8 and 27.5 per cent (Elson 2000). Changes in employment status, from independent cultivator to unpaid family worker with the expansion of cash cropping in Africa, from independent cultivator to wage labourer in India as landlessness increases, and from permanent plantation worker to wage-earning proletariat in Latin America with the rise of agribusiness, appear to be disproportionate among women workers. Some of these variations in the role of women in national censuses may reflect societal changes in the perception of women's roles.

In general, farms run by women tend to have poorer soil and to be smaller and more isolated than those cultivated by men (Momsen 1988a). The crop/livestock mix on female-operated farms is also different, with the emphasis on production for home use rather than for sale, and where cash production does occur sales are made predominantly in local markets rather than for export. Small animals, such as chickens and pigs, which can be fed on household scraps, are kept more often than cattle. Because women smallholders often find it difficult to hire men to undertake tasks such as land preparation and pesticide application, they may be forced to leave some of their land uncultivated and to concentrate on subsistence production.

It appears that the feminization of agriculture, which seemed to be associated with development in the 1980s, was reversed in the 1990s as new employment opportunities for women, especially for young women, open up in most countries. However, older women are increasingly over-represented in the populations of rural areas. They are unable to take advantage of the new employment opportunities for women in manufacturing and services because of age and lack of education, but as farmers they are not as productive and innovative as younger people, although they preserve traditional knowledge of plants and farming.

Farming systems and gender roles

It has been argued that regional differences in the female contribution to the agricultural sector are related to the ways in which people extract food from the environment. Before we became agriculturalists some 10,000 years ago, we were all hunters and gatherers and it is thought that under this system women produced the major part of the food consumed. Studies of the few remaining hunting and gathering societies, none of which is totally untouched by the modern world, have shown that gathering by women is the major source of food in over half of those societies (Plate 6.5). Women may also participate in communal drives of herds, may be responsible for hunting small animals and collecting insects and reptiles, and may work with men in fishing. Hunting is precarious

Plate 6.5 *Australia: Aborigine woman digging for roots and witchetty grubs in the outback.*
Source: author

and uncertain and so gathering tends to provide the basic diet. In contemporary hunting societies like those of many aboriginal groups in Australia, the nutritious grubs and 'bush tucker' gathered by the women often provide a major part of the food intake, but only the meat which the men procure is socially valued and shared.

Women may have been the first agriculturists, as the step from gathering roots and seeds to planting and cultivating is but a small one. As agriculture has developed, it is possible to recognize male and female farming systems (Boserup 1970). In the extensive, shifting, non-plough agriculture of tropical Africa most of the work in the fields is done by women and Ester Boserup (1970) deemed the system to be female. In the plough culture of Latin America and Arab countries there is low female participation and the system is identified as male. Both men and women are equally involved in the intensive irrigated agriculture of South-east Asia. This typology of farming systems according to gender roles is attractive but easily becomes a form of agricultural determinism. At the local level, relationships between types of farming system and female participation are very complex.

Geographical region appears to be a major explanatory variable in differentiating women's participation in agriculture, but the distribution of land between large and small farms proves to be more important. Areas with many small farms usually have relatively high proportions of women agricultural workers and farmers. Land distribution accounts for 44 per cent of the total variation in the female share of farm labour and no less than 80 per cent of the variation attributable to regional differences. The highest proportions of smallholdings are found in sub-Saharan Africa, Asia and the Caribbean, while Latin America has the lowest.

Gender divisions of labour in agriculture

The particular tasks done on farms by men and women have certain common patterns. In general, men undertake the heavy physical labour of land preparation and jobs which are specific to distant locations, such as livestock herding (Plate 6.6), while women carry out the repetitive, time consuming tasks like weeding, and those

Plate 6.6 *Kashmir: a family of herders drying grain. The woman is spinning wool.*
Source: author

which are located close to home, such as care of the kitchen garden (Momsen 1988a). Women do most of the seed-saving and preservation of germplasm, thus playing a major role in protecting agrobiodiversity in most cultures (Plate 6.8) (Tsegaye 1997; Sachs *et al.* 1997). Men and women often have different types of ethnobotanical knowledge, with women knowing about plants used for healing purposes and men often knowing more about plants used for fish poisons or in rituals (Gollin 1997). Among the Karen of northern Thailand men collected only about one-third of the species collected by women (Johnson 2001).

Gender differences in knowledge of uses of plants is also related to gender divisions of labour and gendered spatial use of different ecosystems. Many conservation programmes overlook men's and women's unique knowledge of natural resources that can be vital to project success. For example, in Rwanda agricultural researchers used the knowledge of women farmers to develop new varieties of beans and found that the yields produced by women farmers were consistently higher than those of male farmers, in part because of women's knowledge of the local agro-ecosystem (USAID 2001).

Women generally realize the danger of pesticides to pregnant and nursing mothers and so in most cultures the application of pesticides is considered a male task, and these chemicals are not applied to crops for home consumption (Harry 1980). Irrigation may be used by women but is designed by and for men, and irrigation often increases women's workload by enabling greater intensity of farm production (Lynch 1997; Ramamurthy 1997). Women do a major part of the planting and weeding of crops. Care of livestock is shared, with men looking after the larger animals and women the smaller ones. Children may assist in feeding and herding livestock. Marketing is often seen as a female task, especially in Africa and the Caribbean, although men are most likely to negotiate the sale of export crops. This division of labour is not immutable and may be overridden by cultural taboos.

In many Muslim countries location is more important than the nature of the task in determining the division of labour. In villages in Bangladesh, women only do post-harvest tasks with field crops, which can be done at home, but do all tasks in the kitchen garden, including pest control, with no input from men (Oakley 2002). They do not market crops from fields or gardens as this would involve coming into contact with non-family men. In Turkey, although also

Muslim, the location of fields within the boundaries of the village is redefined as private space so that women can work there even though they are publicly visible (Daley-Ozkizilcik 1993). Turkish women only do weeding for cereal crops, but carry out a full range of tasks, from planting to post-harvest work for cotton, tobacco, fruits and vegetables, where there is less mechanization (ibid.). Men do jobs involving the use of machinery. The average hours worked by women on family farms decrease for those with larger farms, as these wealthier families are able to substitute paid female workers for unpaid family labour. Only in families with the largest holdings are there women who never work on the farm except in a supervisory role. As men take jobs off the farm, women are expected to compensate for this loss of labour, as it is their family duty to do the farm work. The increased income brought in by husbands is not used to pay labour to replace their wives' work. Nor do husbands help by taking over any of their wives' domestic chores (ibid.).

Plate 6.7
Burkina Faso: a woman brewing sorghum beer. She is stirring the second boiling after straining the sorghum grains out. The mixture will then be allowed to cool and left to ferment. Note the use of an improved stove which uses less wood than open fires.

Source: Vincent Dao, University of California, Davis

Between 1955 and 1985 the proportion of the female labour force employed in agriculture in Turkey declined only slightly from 96 per cent to 90 per cent (ibid.). Since then migration to cities and increased education of women has allowed a greater range of jobs for women, mainly for the unmarried, but even in the late 1990s 70 per cent of working Turkish women were still employed in agriculture (World Bank 2001).

Some tasks are gender neutral. In East Africa the production and sale of sorghum and maize beer, and of beer and palm wine in West Africa, is a female task, although men tap the trees for the sap (Plate 6.7). In Indonesia men are responsible for all stages of production of alcoholic beverages and their distribution. The introduction of a new tool may cause a particular job to be reassigned to the opposite sex and men tend to assume tasks that become mechanized. Behaviour in the individual household is often more flexible than the broad general picture suggests because of personal preferences and skills, economic necessity, or the absence of key members of the household.

In areas with high male migration many women become farm operators in their own right. In the eastern Caribbean 35 per cent of small farms are run by women. Table 6.1 shows the gender division of farm work on two islands: Montserrat, with an Afro-Caribbean population and a relatively high proportion of female farmers, and Trinidad, with a majority of East Indian small farmers and very few women farm operators. These differences have resulted in less gender-specificity of tasks in Montserrat than in Trinidad, although farm tasks appear to be shared to a greater extent in Trinidad.

Table 6.1 Gender divisions of labour on small farms in the Caribbean

Task	Trinidad (% of farms)			Montserrat (% of farms)		
	Male	Female	Joint	Male	Female	Joint
Preparation of soil	100	0	0	65	20	15
Planting	72	14	14	42	38	20
Weeding	0	50	50	35	42	23
Pest control	84	6	10	75	25	0
Fertilizing	34	33	33	75	25	0
Harvesting	16	34	50	31	45	24
Care of livestock	14	49	37	53	16	31

Sources: Harry (1980); J. H. Momsen (1973) fieldwork.

Biodiversity

It has been estimated that the net economic benefits of biodiversity are at least US$3 trillion per year, or 11 per cent of the annual world economic output (USAID 2001). The variety and variability of genes, species and ecosystems is a global capital asset with great potential for yielding sustainable benefits. Biodiversity has declined due to habitat destruction, over-harvesting, pollution and the inappropriate introduction of exotic species of plants and animals.

One of the objectives agreed to at the Rio Earth Summit in 1992 was to '[r]ecognize and foster the traditional methods and the knowledge of indigenous people and communities, emphasizing the particular role of women, relevant to the conservation of biological diversity and the sustainable use of biological resources' (Quarrie 1992). Women use wild plants to add to household diets and for herbal remedies, to feed animals and to provide pesticides and compost for their gardens. They often gather these plants from common lands, such as roadsides and forests, but may also transplant them to their household gardens, making them more accessible and also protecting them *in situ*. They are helped by children but boys give up as they get older (Johnson 2001). In a study in northern Thailand, of 43 species identified as declining in availability or disappeared, 17 had been transplanted to home gardens by women, so protecting local biodiversity (ibid.). As wild plants become more difficult to find, women may stop using them as the time and energy needed to look for them becomes too great. Thus only the ones that will grow in household gardens will still be used and conserved. In Burkina Faso wild plants were generally not cultivated in home gardens as this was considered too difficult, but many were dried and on sale in local markets (Smith 1995).

Wild plants provide important micro-nutrients in diets and may be vital to survival in famine periods but knowledge and use is gendered (Figure 6.1). In Burkina Faso wild plants made up 36 per cent of all plants consumed, but people had lost their knowledge of wild famine foods because of improved government assistance at such times (ibid.). In Laos women gathered 141 different types of forest products, including bamboo shoots, rattan, mushrooms and sarsaparilla. One quarter of Laotian women gathered from the forest every day and 75 per cent gathered at least once a week (Sachs 1997). In the Kalahari desert fruits, nuts, gums, berries, roots and bulbs gathered by Kung women contribute 60 per cent of the daily

CONSERVED TRANDITIONS	PRIMARY RESPONSIBLE PARTY	NATIVE TAXA INVOLVED	PLANT PART(S) USED
Wild-harvested foods: plants	♀	*Acrostichum aureum* *Champereia manillensis* *Dioscorea* spp. *Gigantochloa* spp.	fern fronds tree leaves wild yam tubers bamboo shoots
Wild-harvested foods: animals	 ♂	*Artocarpus elasticus* *Schleichera oleosa*	*Artocarpus* latex (trap) Celon oak bark (poison)
Wild-harvested foods: honey	 ♂	*Schleichera oleosa* *Sterculia foetida*	Ceylon oak tree (bee habitat) Javan olive tree (bee habitat)
Wild-harvested foods: drink (Palm wine tapping and distilling)	 ♂	*Arenga pinnata* *Borassus flabellifer* *Momordica charantia*	aren [sugar] palm sap lontar palm sap bitter cucumber fruit (additive)
Weaving: bamboo baskests – small	 ♀	*Gigantochloa* spp. *Grewia* spp. *Pandanus* spp. *Ricinis communis*	bamboo stem vine stem pandan fronds (inner lining) castor oil latex
Weaving: bamboo baskets – large	 ♂	*Dendrocalamus asper*	bamboo stems
Weaving: palm frond baskets	 ♀ ♂	*Corypha utan*	gebang palm fronds
Weaving: pandan mats	♀	*Pandanus* spp. *Ceiba pentandra* *Caesalpinia sappan*	pandan fronds kapok fibers (stuffing) sappanwood bark (dye)
Weaving: building materials	 ♂	*Imperata cylindrica* *Dendrocalamus asper* *Tectonis grandis*	alang-alang grass stems (roof) bamboo stems (frame, walls) teak wood (beams)
Herbal medicine: midwifery	♀	*Ficus* spp. *Mangifera* spp. *Momordica charantia* Musaceae	fig bark mango bar bitter cucumber leaves banana trunks
Herbal medicine: veterinary	 ♀ ♂	*Artocarpus altilis* *Mangifera* spp.	breadfruit leaves (ashes) mango bark
Livestock management: feed	 ♀ ♂	*Artocarpus heterophyllus* *Ficus* spp. Musaceae	jackfruit leaves fig leaves banana trunks
Livestock management: herding (water buffalo)	 ♂	*Entada phaseoloides* *Ficus* spp.	'tarzan' liana vine (cordage) fig vine (cordage)

Figure 6.1 *Gender-differentiated ethnobotanical knowledge in eastern Indonesia (Flores).*

Source: Jeanine Pfeiffer, University of California, Davis

caloric intake and many of the essential vitamins needed by the population (ibid.). Women also harvest small animals, such as the witchetty grubs and goanna caught by Aboriginal women in central Australia.

Women often keep a garden close to the house where they grow special foods that are preferred by their family, such as traditional varieties of vegetables, herbs and spices, thus maintaining biodiversity. In these kitchen gardens they may also grow flowers for ceremonial purposes and plant material for dyes, clothing, baskets and for making drinks. The gardens also provide medicinal herbs, fodder, building materials and fuel. In a Mayan garden in Mexico I noticed a single cotton plant by the door of the house. The young mother told me that she grew it there to provide cotton balls for cleaning her baby.

Home gardens tend to have greater species diversity than cultivated fields, and tropical gardens are the most complex agroforestry systems known. They are multi-functional, acting as aesthetic, recreational and social spaces (Howard-Borjas 2001). The proximity of these gardens to the house means that they are linked to the domestic sphere and so it is usually women who tend them. Most garden produce does not enter the market, the land areas involved are small and plants cultivated are usually traditional varieties known mainly to local people. Thus this gardening tends to be invisible and disparaged as 'minor' or 'supplemental' to agricultural production. Such attitudes in turn lead to an underestimation of the role of gardens in plant conservation. Food grown in gardens is not necessarily supplemental and urban gardens, as in Havana, Cuba, and in the former Soviet Union and Central and Eastern European countries are cultivated intensively and provide a substantial proportion of household food.

Migrants often carry seeds with them and plant them in gardens in their new environment, thus preserving a part of their natal landscape and conserving traditional herbs for both medicinal and culinary uses. Such plant diffusion is almost always done by women as among Hmong refugees in Sacramento, California, where no men had gardens (Corlett 1999). Corlett identified 25 exotic species in these urban gardens. Younger women cultivated on average two garden plots each, growing a total of 26 plant varieties, while older women (over 60 years of age) had three plots and grew an average of 38 plant varieties (ibid.). The older women knew most about the medicinal uses of these plants (ibid.).

Plate 6.8
Bangladesh: a woman displaying some of the seeds she is saving for future planting on the family farm in a north-western village. Women use a variety of traditional ways of both preserving and germinating seeds. Some of the storage containers used are seen next to her. Preservatives are sometimes added to the containers which must then be sealed tight and protected from moisture, insects, rodents and disease. Seeds are often reprocessed midway through the year or during wet seasons.

Source: Emily Oakley, University of California, Davis

In most farming societies women, especially older women, are also in charge of saving seeds for the next season's planting and so make a major contribution to the biodiversity of crop production (Plate 6.8). Seed-saving includes seed selection, processing, storage and exchange. These gender roles mean that women often have a greater ability to recognize varieties of crops. Women differentiate between grain varieties on the basis of the following criteria: grain colour, grain size, time of ripening, taste, cooking quality and time, hardness or softness for grinding, storage life and nutritional quality. Zimmerer (1991), in a study of potato cultivars in the Andes, looked at gender differences in ethnobotanical knowledge. He found that male farmers were less accurate than women when naming species, applied fewer names and incorrectly named uncommon taxa. Women almost exclusively manage potato and maize seed and men are forbidden to handle seed or enter seed-saving areas. As men migrate to take on paid labour the gap between male and female knowledge of plants increases (ibid.).

Women as the predominant managers of plant biodiversity are affected by genetic erosion through the diffusion of modern varieties and the increasing commoditization of plant resources, decreasing access to common land and to cultivable land, and changing consumption patterns. In Bangladesh, population pressure on land and rising water levels have led to women being forced to grow the new high-yielding varieties of rice in order to feed their families adequately and to abandon production of the traditional lower-yielding varieties, that they preferred, because of lack of space. Women's responsibilities for post-harvest processing and family food supplies mean that they endeavour to ensure that varieties grown are acceptable to the household in terms of being palatable and nutritious and that they meet processing and storage requirements. Genetic erosion is therefore tantamount to a form of cultural erosion and loss of social status for women.

Plant breeders of high yielding varieties (HYVs) of crops aim at increasing yields but also need to consider the interests of the women who harvest them and feed their families with them. In parts of Zambia cassava leaves are the main and sometimes the only dark green leafy vegetables available to supply Vitamin A, which prevents night blindness. Women are well acquainted with the local varieties and select them for their palatability and for their ease of harvest. Introduced varieties, promoted for the yield of starch from the roots, are not as acceptable as leaf food, as they grow too tall for women to harvest easily. Some varieties of higher-yielding rice were bred with short stems, the better to support their heavy ears of grain. However, the shorter stems meant women had to bend lower to cut the rice, thus getting more tired, and many families did not like the taste of the new rice. The HYV rice also did not provide the straw for thatching, mat-making and fodder, husks for fuel and leaves for relishes that were important by-products of local varieties.

By 1986 the International Rice Research Institute (IRRI) in the Philippines had become convinced that ignoring or omitting women from the extension and promotion of improved agricultural technology was a mistake and so IRRI set up the programme, 'Women in Rice Farming Systems', to implement gender awareness and training. This has been shown to enhance the success rate of applied and adaptive research.

Women may maintain local varieties of grains and do their own breeding to obtain varieties that have the qualities preferred by their families (Tapia and de la Torre 1997). There is often a strong interest in maintaining and restoring local varieties of crops (Satheesh 2000). Informal seed exchange systems are often female domains and involve mechanisms such as bride price, gift giving and kinship obligations as well as barter and market transactions. The quality of seeds depends largely on women's skills in selecting and storing them.

Gender and agricultural development

The gender impact of the modernization of agriculture is both complex and contradictory. It varies according to the crops produced, the size of farm and the farming system, the economic position of an individual farm family and the political and cultural structure of societies. Men were the favoured recipients of education and technical assistance, and benefited from laws that granted them control over vital resources, most importantly land. The net effect was that women's burden often increased, while men moved into the cash economy (Boserup 1970). Table 6.2 indicates some of the possible gender effects of a number of different changes in rural areas.

Women have often been excluded from agrarian reform and training programmes in new agricultural methods because Western experts have assumed the existence of a pattern of responsibility for agriculture similar to that of their own societies, in which men are the main agricultural decision makers. This error has resulted in failure for many agricultural development projects. Even when included in development projects, women may be unable to obtain new technological inputs because local political and legislative attitudes make women less creditworthy than men. Where both men and women have equal access to modern methods and inputs there is no evidence that either sex is more efficient than the other. The introduction of high-yielding varieties of crops may increase the demand for female labour to weed and plant, while leading to an increase in landlessness by widening the economic divisions between farmers. On the other hand, technological change in post-harvest processing may deprive women of a traditional income-earning task. In Sri Lanka the introduction of high-yielding varieties of rice, and of machinery and chemical fertilizers, eliminated women's work in weeding and winnowing, but because of problems with mechanical transplanters they still worked in planting rice seedlings (Pinnawala 1996). This change has reduced the flow of female seasonal labour migrants from the wet zone to the new dry zone settlements in Sri Lanka during the harvesting and transplanting seasons for paddy rice. Women valued this work because it allowed them to get away from dominant husbands and household chores, to earn money which they rather than their husbands could control, and because they enjoyed the social interaction among the migrant work group (ibid.). Although the relationship is not simple, new technology and crops seem in most places to benefit men rather than women, especially when not accompanied by political change.

The growth of export production has provided new jobs for women in agribusiness. Mexico produces early strawberries and tomatoes for the North American market and women provide most of the labour used in picking, grading and packaging these products. Vegetables and flowers for the European market are now produced in East Africa and Bangladesh in the same way. In several Latin American countries, such as Colombia, Costa Rica and Ecuador, and in the Caribbean, fresh flowers and potted plants are grown and airfreighted to North American and European markets.

In a mere five years Ecuadorean roses exported to the USA have come to generate US$240 million a year and many jobs in a once-impoverished area near Quito, the capital. Ecuador is now the fourth largest producer of roses in the world and the industry employs over 50,000 workers, of whom more than 70 per cent are women. Recent studies have found that these women have an above-average number of miscarriages, and that more than 60 per cent of the workers suffer from headaches, nausea, blurred vision, fatigue and loss of appetite and hair (ibid.). Many of the health problems are the result of the indiscriminate use of highly toxic pesticides, banned in the developed world, with little provision of protective clothing for workers, combined with the stress of working at high speed, cutting, wrapping and boxing flowers. An environmental certification programme has been introduced which is gradually improving working conditions on the flower farms. Most workers earn above the national minimum wage of US$120 a month and 'by employing women, the industry

Table 6.2 The gender impact of agricultural modernization

Changes in the rural economy	Changes in women's socio-economic condition					
	Property ownership	Employment	Decision-making	Status	Level of living and nutrition	Education
I Structural						
Capitalist penetration of traditional rural economy	Loss of rights of usufruct. Increase in landlessness. Sale of small properties.	Proletarianization of labour. Increase in male migration. Increase in employment of young unmarried women in agro-industries, urban domestic employment and multinational industries. Decline in job security. Increase in overall working hours. Triple workload of women as farmers, homemakers and wage labourers.	Increase because of male migration and economic independence of young women.	Increase in proportion of female heads of households because of male migration.	Increased dependence on remittances from migrants and employed children. Loss of usufructuary rights leads to decline in subsistence production and substitution of store-bought goods for home-produced.	Possible increase for daughters as their economic role becomes more important. May decline as increased burden on mothers forces daughters to take on more household tasks.
Land reform and colonization	Women generally not considered in redistribution of land. Loss of inheritance rights.	Decline in women farm operators. Increase in female unpaid family workers.	Decline because of patriarchal nature of colonization authorities.	Decline because of isolation and loss of economic independence.	Increased dependence on male head of household often leads to decline in family nutrition level despite possible increase in disposable income of family.	Increase if improved facilities, decrease if increased physical isolation. Children may be needed to work farm.
II Technical						
New seeds and livestock breeds, pesticides,	May lose usufruct rights as land is used more intensively. Land	Women exclude themselves from use of chemicals because of	Decline. Training in new methods in agriculture may be limited to men.	Increase in family income may allow women to concentrate	New crops may be less acceptable in family diet and nutritionally	Increase in additional disposable income of family may be used for

herbicides, irrigation less accessible to women than to men	owned by women is often physically marginal and not suitable for optimum applications of new inputs.	threat to their reproductive role. New crops may not need traditional labour inputs of women. Women generally displaced from the better-paid, permanent jobs.	Use of new technology and crops generally subsumed by men. Women farmers equally innovative when given opportunity.	on reproductive activities. In patriarchal society this increases status of male head of household.	inferior because of chemicals.	children's education.
Mechanization	Women operate smaller farms in general and so may not find it economic to invest in new implements.	Women usually excluded from use of mechanical equipment. Women farmers have difficulty obtaining male labourers.	Decline.	Decline because of reduced role on farm and downgrading of female skills.	New implements not used for subsistence production.	Growth of interest in mechanical training but limited to males.
Commercialization of agriculture and changes in crop patterns	Female-operated farms tend to concentrate on subsistence crops and crops for local market, and to remain at small scale.	Decline because technical inputs substituted for female labour.	Decline because less involved in major crop production activity.	Decline.	Decline because cash crops take over land traditionally used for subsistence production by women. Male allocates more income to developing enterprise and for personal gratification than to family maintenance.	Increased time available for education.
Post-harvest technology	New equipment owned by men.	Women's traditional food processing skills no longer in demand. May employ young women in unskilled jobs in agro-industries.	Decline because ownership of equipment and skills passed to men.	Decline because female skills downgraded, although women still important as seed savers except where GM seeds introduced.	Decline because loss of women's independent income from food processing activities. New product may be nutritionally inferior. Women deprived of use of waste products for animal feed and so lose important part of traditional family diet.	Increased time available for education.

Table 6.2 (continued)

Changes in the rural economy	Changes in women's socio-economic condition					
	Property ownership	Employment	Decision-making	Status	Level of living and nutrition	Education
III Institutional						
Credit institutions	Microcredit loans encourage expansion of female enterprises.	May allow increase in number of paid workers on farm and release of female family labour.	Decline because of patriarchal control of credit.	Decline.	Increase if credit used wisely. May decline catastrophically if land used as collateral and lost to credit institution.	Increased demand for education as agricultural enterprise grows and horizons broaden.
Cooperatives	Closing of cooperative and collective farms in transition countries led to privatization of land and allocation to women and men.	Work on cooperatives often undervalued.	Decline because not included in cooperative decision-making boards.	Decline.	May lead to more commercial crop production and better standard of living.	Increased time available may increase demand for education from women.
Marketing and transport	Ownership of transport facilities generally male but if women maintain marketing role may invest in a vehicle, and hire male drivers.	Decline in traditional role in marketing with decline in production for local sale and increased size of market area. Difficult for women to travel long distances because of time demands of family and physical dangers.	Generally excluded from marketing decisions as community production is incorporated in national and international system.	Decline because of loss of traditional role.	Decline because of loss of women's income from marketing. Exposure to imported, manufactured goods reduces income from artisanal work such as pottery and weaving. Use of imported foods may reduce nutritional level but may also reduce time spent by women in food preparation.	Need for numeracy and literacy may lead to increased appreciation of advantages of education. Time released from household and marketing duties may be used for education.

has fostered a social revolution in which mothers and wives have more control over their families' spending, especially on schooling for their children' (ibid.: A27).

In Brazil, Mozambique and Sri Lanka cashew nuts are produced for export and women process the crop. This is an unpleasant job involving removing the nut from its protective outer casing which contains an acid harmful to the skin. In north-eastern Brazil women work in cashew-processing factories with no protective clothing supplied. Not surprisingly the workforce has a very high turnover rate. The workers in Brazil are predominantly young, unmarried women, but in Sri Lanka they tend to be older, married women. In some villages in Sri Lanka these women have banded together and obtained adequate credit to enable them to set up as petty commodity producers of cashews (Casinader *et al.* 1987). In this way they eliminate the share of the profit which used to go to middlemen and have a home-based, income-earning occupation which can be integrated with household tasks and childcare. The additional income accruing to these women has given them a stronger voice in community affairs and changed the traditional patriarchal power balance in many households. Once again we have an example of the differential and unpredictable impact of change on women.

Gender differences in time budgets

Women on small farms in the Third World often have a triple burden of work. They are expected to carry out the social reproduction of the household, which may include long treks to fetch water and firewood, wearisome journeys to take children to a distant clinic, time-consuming preparation of traditional food and the depressing job of trying to keep poor-quality housing clean. At the same time, rural women usually have to provide unpaid labour on the family farm and to earn money by working on another farm or by selling surplus produce. The combination of productive and reproductive activities leads to long hours for female farmers, making them 'probably the busiest people in the world' (FAO 1993: 37).

Time use studies reveal daily, weekly and seasonal fluctuations in the demand for labour and clarify the trade-off between productive work, household maintenance and leisure at different times of year and in various farming systems (Figure 6.2). They also make it possible to identify age, sex and season-specific labour constraints which may need to be overcome if a new project is to be successful. In most

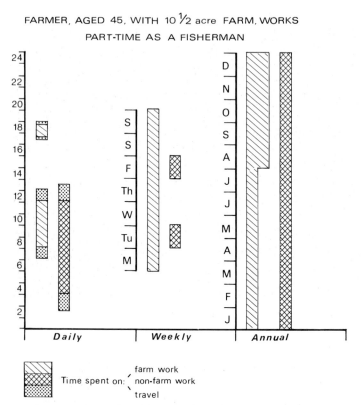

FARMER, AGED 45, WITH 10 ½ acre FARM, WORKS PART-TIME AS A FISHERMAN

Time spent on: farm work / non-farm work / travel

Figure 6.2 *Daily, weekly and yearly time use patterns of a farmer in Nevis, West Indies. He is 45, has a farm of ten-and-a-half acres and works part-time as a fisherman.*

Source: author's fieldwork

rural communities women work longer hours than men and have less leisure time. In the Gambia women spend 159 days per year in work on the farm, while men spend only 103 days a year in farmwork; the women also spend an additional four hours per day on household maintenance and childcare. In Trinidad it was found that women worked longer hours than men on rice and dairy farms but less than men on sugar cane holdings.

When there is a labour shortage at busy times of the agricultural year women will often be expected to sacrifice their remaining leisure time for additional farmwork, acting as a reserve labour force. In Nevis, in the West Indies, for example, weekly hours worked by women on the farm equal 72 per cent of male hours in the busy season but only 66 per cent in the less busy period of the agricultural year. In Sri Lanka, on the other hand, women work longer hours than men in the peak season and only very slightly less in the slack season. Despite their major contribution to agricultural production, women still do almost all the housework and the collecting of wood

Plate 6.9 *Sri Lanka: a woman watering vegetables near Kandy.*
Source: author

and water, and are responsible for most social and religious duties.
Consequently they have much less time than men for leisure
throughout the year and at the peak agricultural season sacrifice an
hour a day of their sleep and leisure time for extra farmwork, despite
reducing the time they spend on their reproductive tasks. Thus
women's workhours in rural areas of the dry zone of Sri Lanka
average over 18 hours per day in the peak season compared to 14 for
men. In the slack dry season agricultural workhours decline more for
women than for men but women compensate by spending more time
cleaning, cooking and collecting water. Both women and men spend
more time on social/ceremonial and leisure activities and on sleep in
the dry season but women's total monthly workhours fall only from
560 to 530, while men's go from 426 to 350 (Wickramasinghe
1993). Over 65 per cent of production activities are done by women
in dryland farming, compared to 30 per cent of the tasks in paddy
rice-growing areas (ibid.) (Plate 6.9).

In Zambia today women do 60 per cent of the agricultural work but
domestic chores take twice as long as farmwork. Food preparation is
one of the most onerous tasks and development has had little
beneficial effect on the time and energy needed for domestic labour.
The heavy burden of both productive and reproductive work has

contributed to a high incidence of ill health among rural Zambian women and to poor family nutrition. Over 50 years ago, during a period of high male migration, it was noticed that, at times of heavy demand for agricultural labour, women frequently failed to cook meals and the family went hungry, despite plentiful supplies of food, as the women were too exhausted to collect firewood and water and to gather the relish needed for the meal. Contemporary research suggests that the recent intensification of women's labour input on farms as a consequence of the introduction of maize cash-cropping may have contributed to an apparent increase in child malnutrition, because women have less time to prepare meals, especially weaning foods, for their families.

Lack of time is resulting in a de-intensification of farming in central Java. Increased need for regular reliable cash income is pushing both men and women farmers into paid non-farm employment and poor women farmers in particular find that the combined demands of reproduction and non-agricultural production leave little time for farmwork. In a study in Burkina Faso, although shortage of time for rural women was reducing the use of wild plants, the women did not necessarily wish for more help from men (Smith 1995). It was found that 36 per cent of foods consumed came from wild plants and women and girls collected over 80 per cent of the volume of wild plants. The females mainly collected leaves for use in sauces for the family, while the men and boys mainly collected fruits for their personal consumption (ibid.). Where gendered knowledge was less specific, as in cultivation, collecting wood and fetching water, women would have liked more assistance (Table 6.3).

Table 6.3 Gender roles and time use in rural Burkina Faso

Activity	Percentage of women's work hours	Actual male assistance[a]	Desired male assistance[a]
Cultivation	26	100	52
Collecting wood	23	57	44
Fetching water	18	9	22
Preparing meals	18	0	4
Searching for wild edible plants	15	0	0

Note: a percentage of focus groups (N=23, made up of 195 informants of whom 72 per cent were women) responding that men helped them in these activities.

Source: adapted from Smith (1995).

Plate 6.10
St Lucia, West Indies: women loading bananas for export to Britain. This job has now been mechanized and is mainly done by men. The loss of a protected market in Europe under new WTO rules is leading to a rapid decline in West Indian banana production from both plantations and small farms.
Source: author.

Women in the plantation sector

Plantations are the organizations which enabled the integration of many parts of colonial empires into the periphery of the world economy. The archetypal example is the sugar plantation in the Caribbean and Brazil in the seventeenth century, but plantations were also established at a later date, for rubber in Malaysia and tea in Assam and Sri Lanka. Female labour has been very important in this sector. Under slavery, women came to form an ever greater proportion of the field labour force on Caribbean sugar plantations and at emancipation were in a majority. Plantation labour is a declining source of income for poor women in rural areas of the Caribbean (see Plate 6.10). However, in Malaysia women account for over half the plantation labour force and this proportion is increasing.

The importance of women as plantation labourers today reflects a decline in available male labour and the lack of alternative employment opportunities for rural women.

Work on plantations is generally unskilled, poorly paid and seasonal (see Box 6.1). In many cases, the only way for the household to survive is for women and children, as well as men, to work as a unit

Box 6.1

Gender on tea plantations in Sri Lanka

Tamil workers from India were brought to Sri Lanka by the British in the nineteenth century to work on the newly developed tea, rubber and coconut plantations. They account for only a small proportion of the total population of the country but they are concentrated on the tea plantations of the central part of Sri Lanka. Among the Tamil plantation workers the female participation rate in the workforce is 54 per cent compared to only 17 per cent for rural non-estate women. However, the physical quality of life of these women workers is much inferior to the national average. The Tamil plantation workers have above average maternal and infant mortality rates, very low fertility rates, high female illiteracy rates and are the only group in Sri Lanka in which women have a lower life expectancy than men. The group's general poverty has been exacerbated for women by patriarchal norms which have resulted in women's subordination and reduced their access to basic needs.

The female tea plucker's day begins before sunrise. She gets up at around 4.00 a.m. to prepare breakfast and lunch, clean the house and get the children ready for crèche and/or school. The morning meal consists of homemade bread (*roti*) with a watery curry and tea. By 7.00 a.m. the tea pluckers are at work in groups and keep filling their baskets with leaves until the tea break from 9.30 to 10.00 a.m. The lactating mothers visit the crèche to nurse their babies and then resume work until 12.30 or 1.00 p.m. The woman worker takes the load to the weighing shed, visits the crèche, nurses her baby and goes home for the midday meal, which she has prepared the night before. She returns to the field by 2.00 p.m. and continues to pluck leaves until 4.30 p.m. She takes the load to the weighing shed and waits her turn. She visits the crèche to collect the children and returns home at around 5.30 p.m. She then starts the evening chores: cleaning the house, preparing the evening meal and the next day's midday meal, feeding the children, cleaning them, washing the clothes and putting the children to bed. She is often the last to go to bed at around 10.00 or 10.30 p.m. She sleeps on a sack on the floor as there is usually only one cot in the one-roomed house, which is used by her husband.

The per capita calorie intake of plantation workers is one of the highest in Sri Lanka but 60 per cent of the workers suffer from chronic malnutrition, at a rate which is almost twice as high as in the rest of the population. Part of the explanation for such apparent dietary deficiencies seems to lie in the subordinate position of women in the family, their heavy workload and their lack of time for food preparation. The woman tea picker's work involves climbing steep slopes carrying a weight of up to 25 kilos in a basket on her back and constant exposure to rain, hot sun and chill winds. She has neither time nor energy to prepare nutritious meals for her family. Nor does she have time to purchase a variety of foodstuffs as the plantation store is expensive and has few items in stock, while other shops are too far away to be visited frequently. As meat and fish cannot be stored because of lack of refrigeration, these items are eaten rarely. Until 1984 female workers, although working longer hours than male plantation workers, received lower wages. They now earn equal daily wages but women rarely have time to queue to collect their wages and so their husbands collect them for them. In many cases the wages are spent on alcohol and gambling. Even maternity payments are collected by husbands and so this money is rarely spent on supplementary food for the lactating mother or newborn baby. In the mid 1990s, 28 per cent of estate workers were living below the poverty line but food stamps and other welfare benefits were not available to them (Manikam 1995).

Plantation workers have strong unions but, although more than half the workers are female, there are no women union leaders. Better access to basic needs, such as improved schooling, widespread health care and more sanitary living conditions seem to receive low priority in union demands. Household technology remains primitive and time-consuming, and the allocation of time between increased wage work and household chores is a constant balancing act for women plantation workers. In the mid 1990s Manikam found that only 2 per cent of plantation houses were in good repair, 76 per cent needed major repairs, and 4 per cent were so dilapidated they had to be demolished (ibid.). And in a survey in 1986/7 it was found that 65 per cent of urban houses had electricity, as opposed to 20 per cent of rural houses and only four per cent of houses on tea estates (ibid.). By the 1990s employment on tea estates was declining and the children of estate workers needed to be educated for different jobs. Despite foreign aid contributions Tamil medium schools in plantation areas had fewer trained teachers, worse teacher/pupil ratios and lower standards than elsewhere in the country (ibid.). A Social Welfare programme was instituted on the plantations in 1985 to deal with many of these problems, especially health facilities, water supplies and housing, but after privatization in 1993 progress slowed.

Source: Samarasinghe (1993) and Manikam (1995).

to maximize output. This was the traditional method on Brazilian coffee plantations for harvesting the beans and is used on Malaysian rubber plantations. In both cases the estate provides a small subsistence plot for the family and expects the women and children to act as a reserve pool of labour which can be called on at peak periods of demand. Such an employment pattern means that women are encouraged to have many children and cannot afford to allow these children to take time away from plantation labour to attend school.

Women as rural traders

Not only do women produce, process, prepare and preserve agricultural products but they are responsible for much of the trade in these, and other goods, in many parts of the South. The sale in the local market of surplus production is an important element in small-scale agriculture. In the Caribbean this role developed under slavery and was seen as so important to the local economy that the plantocracy allowed women traders the freedom, despite their slave status, to travel around within each island. Thus women had a powerful position as carriers of news and information between slave plantations. In many parts of the world, women continue to play an important role as rural information sources and providers of food to

Plate 6.11
South Korea: a woman diver on the south coast of Cheju Island. The women wear wetsuits to dive for shellfish.

Source: Janice Monk, University of Arizona

Plate 6.12 *South Korea: women divers preparing shellfish by entrance to a restaurant on Cheju Island.*

Source: author

urban areas. This may involve food from the sea as well as from the land. Although women rarely work as fisherpeople they are often involved in net-making and the preparation and sale of the catch. Because of the long absences of men at sea, women in fishing communities may be more powerful than in non-fishing communities (Norr and Norr 1997). In South Korea, on the island of Cheju, women dive close to shore for shellfish and crustaceans, while men fish in deeper water from boats (Plate 6.11). The women dive as deep as six metres for eight hours a day for half the year. Since the sixteenth century women have dominated diving for marine resources in Cheju Island. The women work as a cooperative for marketing and make a reliable living from these shellfish, which fetch a good price on the mainland and are served fresh to tourists at nearby restaurants (Plate 6.12). The women of Cheju island have more autonomy than mainland Korean women because of their long history of financial independence. In many cases it is women, especially in West Africa, who through the trading of goods act as a major link and source of information between rural and urban sectors of the economy.

Box 6.2

Rural entrepreneurship in an eastern Hungarian village: a preferred or forced occupation?

Mrs A. runs a flower shop in a village of 2,000 people of whom 500 are poor gypsies. She studied flower arranging for two years in vocational school but worked on the local agricultural cooperative as a store keeper. In 1988 she started her flower business part-time but in 1992, when the cooperative closed down, she opened her flower shop at the age of 47. She was too young to retire, which women could then do at 55, so started her flower business in order keep paying social security so that she would be eligible for a pension. She could have got a job in the county seat 50 km away but felt too old to start commuting. Her husband retired early on a disability pension. She has two adult children and is proud that no member of the family has ever been unemployed.

She and her husband obtained some land from the cooperative and rent it to a large farm. In return they get 2.5 tonnes of maize, wheat and barley each year. They grind this and use it to feed pigs. They keep four pigs in their 330 square metres of garden and also grow vegetables and fruit. They eat three of the pigs themselves, sell one in the village and use the money to buy another four pigs. Her son takes her to the regional capital, a round trip of about 300 km, each week to buy flowers, mainly carnations. She makes wreaths and sells pot plants as well as fresh and artificial flower bouquets. She makes a 30 per cent profit on the flowers bought wholesale. The shop is attached to the house so they can deduct household expenses, such as water, heating, electricity and telephone from the business receipts. She does not make much profit from the shop and cannot expand her business as the village is too small. Despite her insistence that she is only running the shop in order to be able to retire on a pension, Mrs A. clearly enjoys being her own boss and chatting to customers.

Her husband helps with making wreaths, looks after the pigs and does the housework and the cooking. Mrs A. works with him in the garden. They did not take out a loan because her mother lost her house after the Second World War as she could not pay her mortgage, so all the family is afraid of owing money. If they need extra money they sell some of the peppers they grow in the garden. There is no family history of running a business. Her son lives in the village and works as a car mechanic and her son-in-law is a chemical engineer in Budapest and has travelled widely, unlike Mrs A.'s family. Her daughter's mother-in-law is the leader of the Women's Association in the county seat.

At the other end of the village is the most successful entrepreneurial couple. Mr B. had worked as a private carpenter, like his father. The father retired in 1982 but the son hurt his hand in 1983 so he retired on a disability pension and started a business mixing whitewash. Mr and Mrs B. were married in 1977, when he was 23 and she was 20, and have three children. Mrs B. had trained as a cook. She worked in the state restaurant in the village but this closed in 1986. Her husband had the idea of her starting an animal feed mixing business as the one serving the village was too far away for old people to carry the feed. Her husband told her that mixing feed was no different from making a cake, except in scale! Today she sells feed to seven villages and makes deliveries in two large vans. She has bought a shop from a former cooperative farm in another village and has four male employees. In the late 1990s Mr and Mrs B. bought an old house across the road and were building a modern sty for 200 pigs on the land. They have a contract with the sausage factory in a nearby town. They started the piggery for their sons. Mrs B. does all the housework and childcare and her husband does the business paperwork. Their house is equipped with modern labour-saving items, such as a washing machine and dishwasher.

Her parents worked on the cooperative farm where her father was a tractor driver. Mrs B. feels that she would not have become an entrepreneur if her husband had not pushed her into it as 'he did not want me to have any other boss but him!' However, she was clearly very much in charge of her feed business and was proud of her success. They have never taken any loans and have always used their own capital for business expansion. Although they have relatives in Germany and Romania, all their customers are local. They work hard and never take holidays but feel that other villagers are jealous of their achievements.

Source: author's fieldwork, 1997 and 1998.

Women as rural entrepreneurs

In rural areas women are increasingly involved in agroprocessing (Casinader *et al.* 1987), in producing craftwork for export (Momsen 2001) and in doing piecework for city-based manufacturers. These new jobs have been made possible by improved rural–urban links in terms of both transport and electronic communication. Globalization has opened distant markets to local producers (ibid.), yet rural areas still lack many commercial services. In Hungary, women who are spatially entrapped in rural areas by their household responsibilities have led the way in providing new, post-1989, village retail outlets, especially restaurants, shops, hairdressers and bed and breakfast inns (see Box 6.2 and Table 6.4). Women dentists in rural western Hungary have built up lucrative practices in health tourism, serving patients coming mainly from Austria and Germany (Momsen 2002b). Such initiatives are less available to rural populations in eastern Hungary that are far from rich consumers. As a result, people in rural areas in western Hungary were more optimistic than in eastern Hungary and presumably cross-border relationships will improve after Hungary becomes a member of the European Union in 2004.

Plate 6.13 *Romania: a woman in her backyard greenhouse in Curtici, where she grows vegetables for the local market.*

Source: Margareta Lelea, University of California, Davis

In western Romania, although much poorer than Hungary, there is optimism among entrepreneurs, especially those producing foodstuffs for the local market (Plate 6.13).

Becoming an entrepreneur is often a frightening step for people who grew up under communism, although individual activity, mainly in the form of food production on private plots, was more widespread in Hungary than in other Eastern European countries before 1989. Today, most entrepreneurs explain their choice of occupation as forced upon them for household survival, or as a way of reducing taxes, while only a few admit to a desire to be self-employed. Those who are most successful often have a family background in self-employment and have more social capital, especially in terms of contacts outside the country. Young men usually respond to lack of employment in rural areas by migrating nationally or internationally. Or they may opt for early retirement on disability pensions if older. Single women may also move to cities to work in factories or domestic service, but many women are also creating more diverse rural economies through their entrepreneurial activities.

Table 6.4 **Types of entrepreneurial activity in rural western and eastern Hungary, by gender (%)**

Activity type	West		East	
	Women (N = 125)	Men (N = 125)	Women (N = 53)	Men (N = 48)
Agriculture, forestry, fishing, hunting	5.1	13.6	11.4	28.4
Manufacturing, craftwork	13.2	36.9	5.3	14.2
Services:				
retail sales	41.0	14.0	62.1	31.4
personal	18.8	0.2	9.1	0.5
professional	13.3	9.5	6.8	11.3
tourism	2.6	1.0	–	–
transport	0.4	11.4	1.5	5.4
security	–	0.7	–	1.5

Note: totals do not always sum to 100 per cent because of rounding.

Source: author's fieldwork 1998–9.

Learning outcomes

- An understanding of the importance of culture and changing technology, as well as farming systems, in assigning particular roles to men and women.
- An appreciation of how hard rural women work and how they increasingly sacrifice their leisure to take on income-earning activities.
- Recognition that new types of rural employment created by growth of industry and entrepreneurial activity in the countryside are providing new opportunities for some people, especially young women.

Discussion questions

1   Examine how the extent of women's involvement in farming varies according to the nature of local farming systems.
2   Explain how all farming systems have gender-specific agricultural tasks.
3   In what ways is entrepreneurial activity producing a more diverse rural economy?
4   Why does technological change tend to reduce the role of women in agriculture?

Further reading

Awumbila, Mariama and Janet H. Momsen, (1995) 'Gender and the environment: women's time use as a measure of environmental change', *Global Environmental Change* 5 (4): 337–46. Provides comparative examples from West Africa and the Caribbean of changing pressures on women's time.

Boserup, Ester (1970) *Women's Role in Economic Development*, London: Allen and Unwin. The first book on women and development. It provided the stimulus for much of the research in this field over the last three decades. Boserup identifies gender role differences at a continental scale and relates them to types of farming.

Momsen, Janet H. and Janet Townsend (eds) (1988) *Geography of Gender in the Third World*, London: Hutchinson; New York: State University of New York Press. The first edited volume on the geography of women and development. It contains 19 case study chapters on gender issues in developing market economies.

Momsen, Janet H. and Vivian Kinnaird (eds) (1993) *Different Places, Different Voices: Gender and Development in Africa, Asia and Latin America*, London and New York: Routledge. An edited volume containing 20 case studies from all parts of the South, mostly written by people from the area. The editorial introduction emphasizes how authorial voice varies from place to place and so influences our understanding of local problems.

Websites

www.fao.org/WAICENT/FAOINRO/SUSTDEV/WPdirect/default.htm   Food and Agriculture Organization of the United Nations (FAO), an agency that aims to promote agricultural development and food security. This site includes on-line analysis papers on rural gender issues, as well as current activities on gender and agriculture. The Statistics Division of the Economic and Social Department of FAO provides a range of data relating to gender at the FAOSTAT website under the Population Domain.

In countries in the South, the proportion of the labour force in nonagricultural jobs is estimated to have increased from 27.4 per cent of the labour force in 1960 to 40.9 per cent in 1980. There are no more recent comparable global figures available but agricultural employment still occupies the largest proportion of women in most of Africa and Asia (World Bank 2001). In 1990 economically active women in large cities worked mainly in professional, clerical and service occupations: 82 per cent in Latin America, 74 per cent in Africa and 64 per cent in Asia (United Nations 1995a). Fewer women than men worked in production jobs in cities, with only 33 per cent of urban women in Asia and 20 per cent elsewhere employed thus in 1990 (ibid.).

There is no clear relationship between higher levels of development, urbanization and increased female employment. Indeed, in India and parts of Africa, urban growth is associated with a decline in overall female labour force participation. Under-representation of women among nonagricultural employees tends to be greatest in the least developed countries, indicating a time-lag effect on the employment of women in modern occupations.

Female marginalization

It is often suggested that women's role in production becomes progressively less central and important during capitalist industrialization in developing countries. There are four dimensions of 'marginalization' as it is applied to urban female employment. First, women are prevented from entering certain types of employment, usually on the grounds of physical weakness, moral danger or lack of facilities for women workers. Second, marginalization can be seen as 'concentration on the periphery of the labour market', where women's employment is predominantly in the informal sector and in the lowest-paid, most insecure jobs. Third, the ratios of workers in particular jobs may become so overwhelmingly female that the jobs themselves become feminized and so of low status. A fourth dimension is marginalization as 'economic inequality'. This aspect refers to the economic distinctions which accompany occupational differentiation, such as low wages, poor working conditions and lack of both fringe benefits and job security in work thought of as 'women's'.

Clearly the concept of marginalization is complex and is often difficult to identify empirically. It is also a relative concept varying from place to place and cannot be used to predict changes over time. However, it remains useful as a descriptive tool. Female marginalization is usually blamed on efforts by employers to minimize labour costs, but historical, cultural and ideological factors are also important. It has been argued that eliminating gender discrimination in occupation and pay could increase not only women's income but also national income. For instance, if gender inequality in the labour market in Latin America were eliminated women's wages would rise by half and national output could increase by 5 per cent (Elson 2000).

In Bolivia by 1999 the gender parity index for education was at par (World Bank 2001). Theoretically women should have suffered only a very minor disadvantage when competing with men for the vast majority of jobs which require only a primary or secondary education. However, as Box 7.1 shows, labour force participation rates were highest among the few women with post-secondary education, followed by illiterate women, while those with secondary education had the lowest participation rate, as also occurred in Colombia at an earlier period (Scott 1986). This suggests that the Latin American labour market for men and women was divided and that barriers for women were not based on skill levels.

Gender divisions of labour

The main explanations for the disadvantaged position of women in urban labour markets fall into three groups based on different theoretical viewpoints: those of neo-classical economics, labour market segmentation and feminism.

Neo-classical economic theory

This assumes that, in competitive conditions, workers are paid according to their productivity. It follows from this assumption that observed male–female differentials in earnings are due either to the lower productivity of women or to market imperfections. This approach also assumes that women have lower levels of education, training and on-the-job experience than men because families tend to allocate household resources to the education of male family members, while expecting the females, as they grow up, to spend their time on housework and childcare for which training is not required. So neo-classical theory explains gender differences in

Box 7.1

Changing roles of rural–urban migrants in Bolivia

While the majority of peasant women living in isolated communities in highland Bolivia continue to endure untold hardship, often burdened with heavy workloads that threaten their life and well-being, the lifestyles of some more fortunate Aymara and Quechua women have changed dramatically over recent years. Access to secondary education has for some resulted in an increase in social and political awareness and heightened aspirations, self-assurance and assertiveness. Even by 1981, a number of these women in the Lake Titicaca region had become highly critical of the freedom enjoyed by men without domestic responsibilities. Women had begun to demand the right to attend community meetings and participate in decision-making regarding matters likely to have a direct effect on their lives.

As the pace of rural–urban migration accelerates, more and more peasant women are moving to urban environments, either with their families or alone for education and professional training purposes. For example, a significant number of women born in the Lake Titicaca region have now become teachers, nurses and policewomen. Others have improved their entrepreneurial potential by becoming fluent in Spanish and learning simple accountancy. Today some of the most successful marketing women travel widely and freely in the course of their work and may wield considerable power and influence as leaders of market syndicates.

Additionally, an ever-increasing number of women are obliged to make family decisions and assume financial responsibility for their households because of the illness or death of their spouse, desertion or the temporary migration of family members to seek work elsewhere. These and other factors have led women to examine their traditional roles and to consider new, often alien, survival strategies in terms of income generation. To such women, membership in a mothers' club or housewives' committee bestows a feeling of belonging and a sense of security. Such groups promote unity and strength of purpose to achieve set goals, such as the installation of a drinking water supply or the procurement of a teacher to initiate adult literacy classes.

In 2000 two-thirds of the female population of Bolivia lived in cities, compared to just over one-third (37 per cent) in 1970, and compared to 36 per cent of men in 1970 and 64 per cent in 2000. Clearly both men and women have been migrating to urban areas but this migration has affected women's economic activity rates more than those of men: male activity rates in cities were 67.7 per cent compared to rural rates of 77 per cent in 2000, while female rates were 36.9 per cent in cities and 32.3 per cent in the countryside. Female-headed households are also 25 per cent more common in cities than in rural areas. Interestingly, for women with more than 13 years of education, activity rates were higher than those of men in rural areas and, in cities, for this most highly educated group, women's rates were not much below those of men at 61 per cent versus 73 per cent. However, the second highest rate for women was among those with less than three years of education.

Sources: Benton (1993) and ECLAC (2002b).

employment in terms of differences in human capital, where women
are disadvantaged because of their family responsibilities, physical
strength, education, training, hours of work, absenteeism and
turnover. However, it has been shown empirically that these
variables can explain only a part of the wage gap between men
and women.

The neo-classical approach has been criticized on the basis of three
of its underlying assumptions. First, it assumes that the gender-based
wage differential can be largely overcome by improving the
education and training of women. Where differences in education
levels are very marked, as in predominantly Muslim countries, this
may have some initial effect. However, in the long run the result
may be to raise the level of education in 'women's jobs' rather than
to decrease pay differentials. A second implicit assumption is that
men and women have equal access to the labour market and compete
on equal terms for job opportunities. This ignores the gender-based
segregation of the labour market which exists in all countries and
does not appear to decline as gender differences in education levels
even out. It is also affected by cultural issues, as in Latin America
and many Muslim countries, where women are not allowed to work
without their husband's permission (Scott 1986; Daley-Ozkizilcik
1993; Ismail 1999b). A third underlying assumption is that women's
labour force participation is of necessity intermittent because of their
'natural' childbearing role. Yet only pregnancy and breastfeeding are
biologically restricted to women and in most countries of the South
women are able to share childcare with relatives or friends, employ
domestic servants, keep children with them while they work, or they
have access to a free crèche. Sometimes, however, children have to
be left in the care of older siblings or locked in the house while the
mother is at work. In such situations allowing children to work with
the mother has advantages if the family is unable to send their
children to school.

Theories of labour market segmentation

This approach emphasizes the structure of the labour market in
explaining sex inequalities in employment. It assumes that the labour
market is segmented by institutional barriers, but within each
segment neoclassical principles still apply.

One such division of the labour market is into primary and secondary
sectors. Primary jobs are those with relatively good prospects of

promotion, on-the-job training and pay, while secondary sector jobs are poorly paid and have little security. Because of the perceived higher turnover of women they are more likely to be recruited into secondary sector employment, while men will be sought for primary sector jobs. Yet turnover and absenteeism are higher in low-level, boring, dead-end jobs, such as those of the secondary sector where women are concentrated, and so these aspects of employee reliability may be explained by sex differences in type of occupation rather than by inherent characteristics of women. Other factors influencing this gender segregation include the better organization of male workers to defend their skills and income differentials, their resistance to competition from cheaper (often female) labour and the role gender relations and patriarchal ideologies play in the control structure of the firm.

In many parts of the South this differentiation within the capitalist sector is given less emphasis since women tend to be generally excluded from employment in this sector. The industrialization process in developing countries is capital intensive and is dominated by foreign capital and imported technology. This type of industry, often referred to as the 'modern' or 'formal' sector, has a low level of labour absorption and is biased against the employment of women because of their lack of formal educational qualifications, their supposed lower job commitment and because capital-intensive skills tend to be considered 'male' skills. Female employment is concentrated outside this sector in the 'informal' or 'traditional' part of the labour market (see Plate 7.1). The production arrangements in this sector include self-employment, outworking, family enterprise and household service, which offer the flexibility needed by women in combining the demands of their reproductive and productive activities on their time. It also provides flexibility of labour supply for large-scale manufacturers who can subcontract work out to small-scale enterprises at times of peak demand.

However, this model ignores the wide range of technologies that exist in modern industry, some of which, such as light assembly work, discriminate in favour of women. It also ignores the increased demand for women workers created by the expansion of the modern sector in the female-dominated clerical, teaching and nursing occupations. It does not explain the high degree of gender segregation within the informal sector nor the frequent movement of individuals between the formal and informal sectors.

Plate 7.1 *India: women working as building labourers near Gwalior, with men as overseers.*

Source: Janet Townsend, University of Durham

A segmentation of the labour market based on gender may also be observed. The existence of two separate labour markets for men and women tends to restrict women's occupational choices. To the extent that there is an oversupply of candidates for women's jobs this may maintain lower pay levels in this segment of the labour market, while restricting competition within the male-dominated segment and thus keeping wage rates relatively high for men. The sex of the workers may of itself lead to women's jobs being defined as unskilled, while jobs filled by men are defined as skilled.

These economic theories tend to assume that gender roles in society are fixed and are the basis of women's disadvantaged position in the labour market. This can lead to the circular argument that because women are not able to earn as much as men in the workforce it makes economic sense for them to stay at home doing unpaid domestic labour. It is clear that economic theories cannot fully explain gender differences in the labour market and much of the marginalization of women is the result of discrimination based on societal prejudices.

Feminist theories

Feminist theories emphasize the importance of social and cultural factors in restricting women's access to the labour market. These approaches tend to see the interaction between the reproductive and productive roles of women as a key issue rather than a fixed condition. The allocation of housework and childcare to women persists in most societies even though women's participation in the labour market is increasing. Female labour force participation in urban areas affects household composition: families tend to be smaller and there may be a shift away from nuclear families to both extended families and female-headed families. At the same time, domestic help is becoming scarcer and more expensive as alternative formal sector opportunities become available for women. Also, the benefits of education, especially for daughters, are increasing and so children have less time to help their mothers in the home. Consequently, the burden of domestic responsibilities falls ever more heavily on one particular woman in the family. Few poor, developing country homes have the domestic appliances commonly available in industrialized countries and household tasks are a very heavy burden.

As women increase their time spent working outside the home in response to both the new employment opportunities and increasing financial pressures, men rarely increase their share of unpaid work in the home. In Mexico, in 1995, 91 per cent of economically active women did unpaid household work for 28 hours a week, while only 62 per cent of economically active men did any unpaid work in the home, spending only an average of 12 hours per week (Elson 2000). In Bangladesh, where women's participation in the garment industry grew rapidly during the 1980s, women still kept responsibility for unpaid family care. Those with the heaviest burden were women in formal work who not only put in more hours per week in paid work than men but also spent more than twice as many hours doing unpaid care work (see Box 7.2). Informal sector male workers did about the same number of hours in unpaid work as their colleagues in the formal sector, but women in the informal sector did less than formal sector workers for a total of almost half the workhours of formal sector workers (ibid.). The extra time spent by formal sector women workers on unpaid household work may reflect their efforts to meet new standards of 'modern' lifestyles. It is perhaps not surprising that Bangladeshi women factory workers are at the forefront of fighting for workers' rights and for their empowerment within the household

Box 7.2

The story of an urban migrant to Dhaka, Bangladesh

My father was a marginal farmer with only five acres of agricultural land so he worked on other people's land as a paid labourer as well. He had no education but my mother had studied up to class four in the primary school. My father's economic situation declined because of floods and cyclones and he lost his paid job. By the time I was 12 years old my father could no longer afford to feed us three children properly. My father's sister had migrated to Dhaka five years earlier and worked as a housemaid. She offered to find me a similar job. I worked for a kind mistress and did light work caring for the baby and washing the baby's clothes. I was taught how to wash clothes using soap powder and how to iron. I was given nice clothes and plenty of food, including items I had never tasted before. I learned how to handle electric appliances, such as the radio, television and freezer, how to answer the telephone and how to cook and serve food properly. My mistress taught me how to read simple stories. When I returned to the village after six months my parents were pleased because I looked like a middle-class girl and because I gave them all my wages. I wanted to stay in the city.

After two years, I was suddenly called home by news that my father was ill. When I arrived in the village I found it was a trick to marry me off to a 25-year-old man with a secondary education and a cattle business. I was forced into this marriage by my parents, but after three months I found out that my husband had TB. I left my husband and went back to Dhaka but I had been replaced in my old job. My former employer helped me to find a job in a garment factory and allowed me to live in her house. Meanwhile, I heard that my husband had drowned when a ferry sank and eventually I remarried. My second husband worked in the factory and, like my first, had completed secondary education, whereas I did not finish primary school.

Within six months of our marriage my husband lost his job and has not been able to find another job that he feels is suitable to his secondary school training. I became pregnant and gave birth to a girl but had to go back to work in the factory to support my family. I am grateful to my former mistress who helped me in adapting to city life, gave me assistance and showed me where I could get free treatment when I was pregnant and even free milk for my daughter. I am thankful to God to get such a kind-hearted mistress. I also think that I made a wise decision to go into household work, which really exposed me to the urban world and helped me to become what I am now. At the same time, I blame myself for my present situation, in which I am struggling with poverty and living in one room in a slum for which I have to pay 40 per cent of my salary.

Source: adapted from Huq-Hussain (1996: 193–6).

(Hussain 1996). Women's handicap in the labour market because of domestic responsibilities may be growing rather than diminishing in many cities in the South. At the same time as time pressures for poor women are increasing, better-educated women are benefiting from the opening up of well-paid professional, administrative and managerial positions to women. Such women turn to immigrant paid domestic help to assist with carework in the home but this in turn brings its burden of guilt and dependence (Tam 1999). These changes in work opportunities for women are increasing the polarization between the well-educated, middle- and upper-class women and poor women.

Sexual harassment may be an even greater problem, or at least a more open problem, in developing countries and in post-communist countries than in developed ones. In traditional societies, a woman who moves out of her accepted family role in order to take a job may be seen as a 'loose' woman. Men who are not used to meeting women in a work situation may fall back on gender-based social expectations and treat their workplace female colleagues as sexually available. Men in supervisory positions may demand sexual favours in return for job security and this may contribute to high turnover rates for women workers. Those women most in need of paid employment may be victimized by sexual harassment, as the option of resignation from the job, which may be their only means of escape, is often not open to them. The ghettoization of women into certain sectors of the economy may be encouraged by fathers who want to protect their young daughters' reputations, while at the same time needing to send them out to work to contribute to the family income. In post-communist countries the sudden appearance of competition for jobs has resulted in discrimination against women.

Non-economic exclusionary measures are sometimes political and legal but most often are based on familial ideology and are sanctioned by informal controls, such as gossip or ridicule. Employment of women in occupations like teaching and nursing is seen as an extension of their domestic role and so tends to be devalued. In many jobs qualities attributed to men, such as physical strength, are valued more highly than those characteristics thought of as female, such as manual dexterity and docility.

Barriers to women's participation in the urban modern sector

Clearly, certain aspects of social, economic and cultural norms determine women's ability to participate in urban employment in developing countries. Modern industry is spatially separated from the home and involves a standard fixed pattern of working hours. Both characteristics cause problems for women with children. In industrialized societies, in recent years, women with family responsibilities have sought a solution in part-time work but this is generally discouraged by employers in the developing world and in former communist countries. Furthermore, daily working hours are often longer and paid holidays shorter in the South, if such benefits even exist. Thus many women put together multiple self-employed occupations in order to gain an adequate income or seek work in the informal sector because of its flexibility. However, the relatively high participation rate in the modern sector of women with post-secondary schooling indicates that women with well-paid jobs are able to cope with the demands of such work because of the availability of cheap domestic help (Momsen 1999). In Sri Lanka the proportion of women employed in professional and technical jobs rose between 1985 and 1995 from 18 to 49 per cent and in Bangladesh between 1990 and 1996 from 23 to 35 per cent (ESCAP 1999).

The burden of domestic work bears most heavily on poor women. They are usually forced to depend on their own mothers, female friends or older children for assistance with childcare. A few countries do attempt to provide workplace or government funded crèches. Only Cuba appears to be ideologically committed in its Family Code to reducing women's double burden of productive and reproductive work by expecting husbands to undertake a fair share of household chores, but this legislation has yet to become fully effective.

Protective legislation applicable to modern industry, such as workhours and maternity leave, may limit women's work opportunities by raising the cost of female labour. Women are under-represented in trade unions and they do not generally hold positions of office, so it is not surprising that issues concerning women are rarely taken up.

There is considerable evidence of employer discrimination against women. Sometimes this is justified by the employer on the grounds of perceived lower productivity and higher absenteeism and turnover of women. The evidence to support these perceptions is not clear and varies from place to place. Where problems can be noted they are generally related to the family responsibilities of women.

Many employers have a preconceived idea of types of jobs suitable for women. They consider only a very narrow range of jobs as open to women and in this way women's opportunities are more restricted than those of men. Lack of physical strength and inability to supervise are the main reasons given for restricting jobs for women. The advantages women offer are seen as related to a willingness to work for lower wages than men and to their greater docility. Employers also think women are most suited to jobs using so-called household skills or where femininity is an advantage, as in the case of waitressing. These stereotyped views limit employment opportunities for women.

Differences in education level also hinder women from entry into the best-paid jobs. However, this may be a self-fulfilling situation for, where it is perceived by parents that the best jobs go only to educated males, it may be thought that investment in a daughter's education is a waste of money. Lower levels of education among women do not explain all the differences in male and female earnings and it may be concluded that equality of education is a necessary but not a sufficient condition for equality of pay. This situation is very clear in Central and Eastern European countries, where both men and women are equally well educated but female-to-male pay ratios are unequal. In the Czech Republic women received only 66 per cent of men's pay in 1987 but this had risen to 81 per cent in 1996; however, in the Russian Federation the ratio fell from 71 per cent of male pay in 1989 to 70 per cent in 1996 (World Bank 2001). A similar mixed picture comes from the Asia Pacific region: between 1990 and 1995 female wages in manufacturing as a proportion of male wages rose from 69 to 91 per cent in the Cook Islands, from 94 to 99 per cent in Turkey, from 88 to 90 per cent in Sri Lanka, from 64 to 71 per cent in Thailand, from 55 to 58 per cent in Singapore, but fell from 69 to 65 per cent in Hong Kong in the face of increasing competition from low Chinese wages (ESCAP 1999; Seager 1997).

The assumption by policy makers that men are the main providers for the family means that, where there is high unemployment, jobs will be found for men before women. When there is a recession

Plate 7.2 *Brazil: weekly market in a small town in Bahia state. Note the horses used as the main form of transport for people coming from the countryside to the market. The Ferris wheel in the background provides the urban entertainment on market day.*

Source: author

women are usually the first to lose their jobs. Most developing countries have higher unemployment rates for women than for men. Yet in cities the proportion of female-headed households is often higher than in rural areas and may be as high as one-third of all households in some urban places. These women may have to support themselves and their children. Single women workers have often been found to contribute more than their brothers to the income of their family. The continuation of the myth that men are able to be the sole breadwinners perpetuates the secondary status of women in the labour market.

It is also in cities that women undertake informal service activities because the large population creates a market and, for scavengers, a source of raw materials. In cities local social control is weaker and rich clients easily available so that prostitution is a widespread occupation, not only for women but also for young boys, men and transvestites. Urban women become street traders selling cooked foods as well as fresh meat and vegetables obtained from farmers, from wholesale markets or from urban farms (Plate 7.2). Today about 200 million urban farmers, men and women, throughout the world, supply food to 700 million people (FAO 2002). This is a very important source of food, especially of meat, vegetables and fruit, which are often absent from the diets of low-income families, but urban agriculture is especially at risk of contamination from sewage in ground water and heavy metals in the air and soil.

Solid waste disposal

One of the by-products of modernization is growth in the use of plastics and other non-biodegradable items and of packaging of goods. This has led to the increasing problem of solid waste disposal. This problem is most acute in densely populated areas, where space for landfills is limited and the production of waste is increasing, such as cities in the South and small tropical islands (Thomas-Hope 1998). However, in such locations the spatial congruence of rich and poor provides an opportunity for the poor to benefit by gleaning from the discards of the rich. In Buenos Aires, in 2002, the economic collapse of Argentina, which has driven many people into poverty, has resulted in an increase in scavengers searching through the bags of rubbish on the streets for food and items to sell. Most of these new scavengers are women and children who live on the poor fringes of Buenos Aires and come into the rich central city to work as hunters and gatherers as part of a new survival strategy. In Ho Chi Minh city in Vietnam women are the main recyclers, going from door to door buying solid waste products from households. Returns are low but it is a main survival strategy for many poor families (Mehra *et al.* 1996).

Where scavenging is long-established so are the gender roles. In 1999, 70 per cent of the urban population in Haiti was unemployed or underemployed and 53 per cent lived below the poverty line (Forbes 1999: 33). In a study of three waste dumps in Port au Prince, the capital of Haiti, it was found that men collect mainly glass and metal for recycling and sale to wholesalers (Noel 2001). The study interviewed 43 scavengers (13 women and 30 men) on three urban dumps. Women collect waste food for feeding their pigs, which are fattened and either used for household consumption or sold. Where males collect food it is usually as young boys assisting their mothers. Women also specialize in scavenging clothes for resale on the streets. They wash and iron the clothes on Friday evening, sell on Saturdays and collect for the rest of the week. This gender specialization is

Table 7.1 *Scavenging in Port au Prince, Haiti, by type of waste collected, gender and age*

Type of waste collected	[Female] N=13				Total (%)	[Male] N=30				Total (%)
	[Age]					[Age]				
	7–14	15–30	31–40	41+		7–14	15–30	31–40	41+	
Food residues	3	–	–	5	62	4	1	2	1	27
Wood	–	1	2	2	39	–	1	3	–	13
Cloth and clothes	–	–	3	1	31	–	1	2	1	13
Metal	–	–	–	–	0	6	12	7	–	83
Bottles	–	–	–	–	0	5	6	5	–	53
Other	–	–	1	–	8	1	3	2	1	23
Total (%)	3(23)	1(8)	6(46)	8(62)		16(53)	24(80)	21(70)	3(10)	

Note: totals do not sum to 100 per cent because many scavengers collect several different types of refuse.

Source: Noel (2001: 53, Table 7).

guided by knowledge of potential buyers and utility of collected products. Only 12 per cent washed waste after collection, probably because of the pressure from buyers at the dump site or the type of waste collected.

Table 7.1 shows that the very young and the old scavenge lighter items, while only men, especially young men, collect metal and bottles, which are sold to middlemen for recycling. Both women and men collect wood, generally for fuel for cooking. One woman collected plastic bags and individual men collected old tyres, iron, pots and cans. The main reason given by 81 per cent of interviewees for specialization was that they knew where to sell the item. Other reasons given by 30 per cent were that it provided more income than other types, 30 per cent because there was less competition for that particular type of waste, 21 per cent because it could be used to feed their pigs, and 7 per cent because they felt more comfortable dealing with a particular type of waste.

Haitian women tend to see scavenging as a long-term source of fuel and pigfeed, while men practise it more as a short-term way of earning money with 37 per cent of men but only 27 per cent of women having worked as scavengers for less than four years (Table 7.2). Men are more likely to be injured while scavenging (Table 7.3), probably because they are dealing with heavier and more dangerous

Table 7.2 Gender differences in years spent scavenging in Port au Prince, Haiti

	Years in scavenging						Total
	1–4	5–9	10–14	15–19	20–24	25+	
Women	3	–	4	3	1	–	11
Men	11	6	5	4	3	1	30
Total	14	6	9	7	4	1	41

Source: adapted from Noel (2001: 56).

objects of glass and metal. Among the men interviewed 70 per cent had suffered injuries, while only 23 per cent of the women had been hurt while working on the dumps. Some 70 per cent of the people interviewed in Haiti said they worked as scavengers because they could not find another job, so it was clearly seen as a survival activity. The price of the scavenged items was fixed by the buyers and the sellers had no possibility of bargaining. The Haitian minimum daily wage since 1995 has been US$1.53. Only 9 per cent of the scavengers made less than US$1.00 per day and five out of the eight in this group were children, while the 23 per cent who were most successful made US$3–5 per day. None of the women were in the lowest bracket, while 30 per cent of the women were in the highest bracket. Those who specialized in recycling metal, all men, made the most money. Those earning in the highest bracket spent less than they earned. Women avoided middlemen by feeding food waste collected to domestically raised pigs and by the direct sale of recycled clothes in local markets.

Another study in Port of Spain, the capital of Trinidad and Tobago, also looked at scavengers (Seelock 2001). This state, in comparison

Table 7.3 Gender differences in occupational injuries among scavengers in Haiti

	Injuries while scavenging		
	Yes	No	Total
Women	3	10	13
Men	21	9	30
Total	24	19	43

Source: Noel (2001: 73).

with the poverty of Haiti, is one of the richest Caribbean countries, with considerable oil and gas resources. This means that items discarded as rubbish are often quite valuable. For example, items collected by women for resale included packets of unused disposable diapers (nappies) discarded by the factory because they were slightly less than perfect. Scavengers in Trinidad earned on average three times the minimum wage and 92 per cent of men, but only 70 per cent of the women, earned above the poverty line. The weekly average earnings of scavengers, known as salvagers or 'human corbeaux' (vultures) in Trinidad, were higher than those of workers in industry and 8 per cent, all men, earned enough to run a car. Rubbish was delivered to the dump 24 hours a day so scavenging could take place at any time. Women's hours spent working at the dump are limited by the demands of their household responsibilities. Men worked longer hours than women: 21 per cent of the men and no women worked over 15 hours a day, while 18 per cent of the women but only 8 per cent of the men worked four hours or less per day. Women tended to specialize in lighter materials, such as rags, diaper seconds or bottles, which they either sold in the market themselves or to specialist buyers. They scavenged items that they did not have to transport to the recycling plant. For many men scavenging was clearly a profitable full-time job (ibid.). Often seen as one of the occupations of lowest status, scavenging can provide a relatively adequate income above the poverty level. The people involved in this occupation are clearly entrepreneurs rather than victims.

Women in cities have to cope with the spatial separation of home and work, often without the support networks of relatives which exist in rural areas. The double burden of production and reproduction has led to both female interdependency as mistress and servant, and to the growth of female support groups to share the burden of family responsibilities. Cities are seen as the locus of modernity and female rural migrants find adaptation to the cities difficult. They often have to live in squatter settlements with no facilities or in crowded dormitories provided by their employer. Adaptation to modern technology and household appliances, and to a greater variety of clothing and food and other aspects of urban living conditions is often achieved by starting out as a domestic servant. A study of perceptions of modernity among migrant domestic workers in Dhaka, Bangladesh, revealed that ideas such as the end of dowries, higher age at marriage, bearing fewer children, and that having a job made a woman complete and independent, were seen as modern by three-quarters of those interviewed (Hussain 1996). Thus despite often difficult living conditions, urban dwellers find it hard to return to their rural roots because of the new concepts about which the city has made them aware.

Learning outcomes

- Employment discrimination based on gender is pervasive and complex.
- Concentration of population in urban areas provides opportunities for illegal occupations, such as scavenging and prostitution.
- Migration to the city helps in the spread of ideas of female empowerment.
- Urban jobs are generally outside the home but in the case of domestic service take place in the employer's home.

Discussion questions

1   What aspects of modernity would be noticed by rural migrants to the city?
2   Why do sexual stereotypes limit women to a narrower range of jobs than is available to men?
3   Discuss why there is a gendered pattern of items selected by garbage scavengers.
4   How does the spatial separation of home and work in cities act as a constraint on women's employment opportunities?
5   Why do cities offer special opportunities for informal and illegal occupations?

Further reading

Anker, R. and C. Hein (eds) *Sex Inequalities in Urban Employment in the Third World*, London: Macmillan. An edited collection providing a broad coverage of gender discrimination in urban employment in several developing countries.

De la Rocha, M. G. (1994) *The Resources of Poverty: Women and Survival in a Mexican City*, Cambridge, MA: Blackwell. A detailed survey of survival strategies of poor women in urban Mexico.

Pearson, Ruth (1998) 'Nimble fingers revisited: reflections on women and third world industrialization in the late twentieth century', in C. Jackson and R. Pearson (eds) *Feminist Visions of Development: Gender Analysis and Policy*, London and New York: Routledge. An overview from one of the major scholars working on women and developing country industrialization.

Potter, R. and Sally Lloyd-Evans (1998) *The City in the Developing World*, Harlow: Longman. Provides an overview of the nature of urban systems and urban dynamics in developing countries.

Thomas-Hope, Elizabeth (ed.) (1998) *Solid Waste Management: Critical Issues for Developing Countries*, Kingston, Jamaica: Canoe Press, University of the West Indies. Brings together 21 papers from a conference. The case studies are mainly from Africa and the Caribbean and focus on the health and environmental problems of solid waste management in developing countries.

Websites

www.ilo.org   International Labour Organization. Founded in 1919 it publishes a comprehensive collection of labour force data.

www.worldbank.org.data   The World Bank Group has 182 member countries and publishes a wide range of data.

8 ▸ Globalization and changing patterns of economic activity

Learning objectives

When you have finished reading this chapter, you should be able to:

- understand the influence of globalization on gender differences in economic activity over the life course
- recognize the new types of employment for women provided by transnational corporations (TNCs), often in Export Processing Zones (EPZs)
- appreciate the problematic nature of microfinance
- be aware of tourism as employer and social catalyst.

The world is more globally integrated than it was in the middle of the twentieth century but it is still far from fully globalized. Globalization processes 'involve not merely the geographical extension of economic activity across national boundaries but also – and more importantly – the functional integration of such internationally dispersed activities' (Dicken 1998: 5). These processes are unevenly distributed, complex and volatile. Technological changes in the production process have allowed it to be fragmented into separate parts, which do not have to be done in the same location. Standardization and increased automation of production have led to a deskilling of work in manufacturing, opening up jobs for less skilled workers. Change in transportation and communications technologies have enabled a new flexibility in the geographical location of the production. Much manufacturing has become 'footloose', moving from country to country in search of the cheapest labour.

These changes have also made leisure travel cheaper so that long-haul holidays, for tourists from rich countries visiting poor countries, form the fastest-growing sector of world tourism. Yet at the local level, transport may still be a problem for women as social norms often make it unacceptable for them to ride bicycles, as for example in parts of East Africa. In most countries far fewer women than men are able to drive and in parts of the Middle East women are not allowed to drive. In Afghanistan driving lessons for women were reinstituted in 2003 but it was not certain that women would be allowed to drive even if they obtained a licence. In many parts of the South the most common way to carry goods is on the heads of women.

Restructuring, as a result of globalization, tends to reinforce and exacerbate existing gender inequalities (Marchand and Runyan 2000). The gender impacts of globalization have been multiple and contradictory and there have been conflicting interactions between local and global economies, cultures and faiths (Afshar and Barrientos 1999). It has been argued (Chang and Ling 2000) that globalization is gendered into two worlds: one is a structurally integrated world of global finance and postmodern individuality largely associated with Western capitalist masculinity; the other is explicitly sexualized and racialized and based on low-waged, low-skilled jobs often done by female migrants for the high-salaried cosmopolitans of the first globalized world. This underside of global restructuring is reinforced by the patriarchal forces of state, religion, culture and family.

There is a new international division of labour (NIDL) associated with the process of globalization, which involves a search for cheap labour and is reinforced by national and international trade agreements and policies. The process of globalization of economic activity is not only strongly gendered but is also spatially linked with urban areas, which are seen as the locus of modernization in developing countries. Yet rural areas, usually considered to be more closely linked to the local than the global, are also becoming more closely integrated with the outside world through migration, improved communications and the growth of multinational agro-industries and mining projects. Such changes are undermining the patriarchal gender contract, under which families are supported by a male breadwinner, as women move into the labour force in response to new employment opportunities and increasing poverty.

Figure 8.1 shows the global pattern of women's participation in the labour force in 1980. Clearly the greatest variation was between countries in the South. In 1980 the lowest participation rates were in Muslim North Africa and the Middle East, plus Ecuador, and the highest in the USSR and sub-Saharan Africa. Figure 8.2 shows that the percentage of women in the labour force has increased everywhere since 1980, with only Iraq, Oman and Saudi Arabia still having less than a quarter of their female population employed (PRB 2002). Increases between 1980 and 2000 were mainly in the Middle East, with declines being recorded in the transition countries and for Cambodia, Bosnia and Malawi (World Bank 2001). The biggest increases in the proportion of women aged 15 to 64 in the labour force between 1980 and 2000 were in Fiji (19 to 38 per cent), Algeria (19 to 31 per cent), Brunei (31 to 52 per cent), Costa Rica (24 to 40 per cent), and the Middle East (Bahrain 17 to 34 per cent, Kuwait 21 to 42 per cent, Oman 8 to 20 per cent, Qatar 13 to 38 per cent, Saudi Arabia 10 to 23 per cent and the United Arab Emirates 16 to 33 per cent) (PRB 2002). It should be noted that some of these increases may be due to measurement changes rather than being real, as in 1980 unpaid family workers and women working in agriculture were less likely to be considered economically active than in 2000.

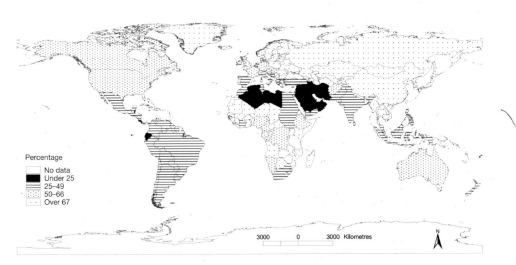

Figure 8.1 *Percentage of women (ages 15–64) in the labour force, 1980.*

Source: Sass and Ashford (2002: 15–20)

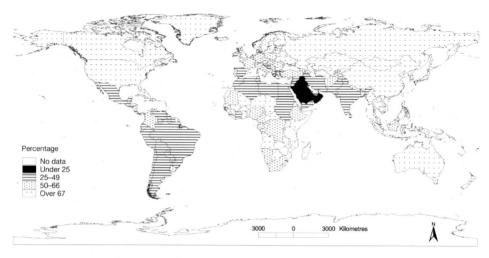

Figure 8.2 *Percentage of women (ages 15–64) in the labour force, 2000.*

Source: Sass and Ashford (2002: 15–20)

In all of the post-communist countries, except Armenia and Slovakia, the proportion of women and men in paid jobs fell during the 1990s with the emergence of officially recognized unemployment (Figures 8.1 and 8.2). In general, the female proportion of the workforce declined, but it increased in the countries in the first group to join the European Union, with the exception of the Baltic countries, and in Bulgaria (World Bank 2001). In Vietnam, where a similar process of reform, moving from a centrally planned to a market economy, has been under way since 1986, the proportion of economically active women increased between 1992/3 and 1997/8, while the proportion of men declined, with more men than women unemployed (Rama 2002).

It has been suggested that the relationship between development and female employment follows a 'U'-shaped curve, with economic activity of women being highest in both least developed and post-industrial societies, while it is lowest in those countries at a middle level of development as women move out of agriculture. Figures 8.1 and 8.2 reinforce this model, showing that Scandinavia and sub-Saharan Africa have similar proportions of women in the labour force, although at opposite ends of the development spectrum. However, at intermediate points cultural, political and historical factors intervene to reduce the applicability of the model.

Socialist countries such as Cuba and China have high participation
rates for women, similar to the levels in North America and Western
Europe (Figures 8.1 and 8.2). Africa has the greatest variation, with
low rates in Muslim North Africa and high rates in parts of Africa
south of the Sahara. Latin America and South Asia have less than
half their female population officially employed, while South-east
Asia has between a half and two-thirds of women employed. The
high rates in Africa are associated with female farming systems but
elsewhere it is urban jobs which dominate. In Latin America women
are employed mostly in the service sector, especially in domestic
service, teaching and clerical occupations. In South-east Asia, the
growth of world market factories employing mostly young women
has led to an increase in female activity rates. For professional
women this double burden has been reduced by the employment of
immigrant women as nannies and domestics (Momsen 1999).

Age also affects the gender division of labour. In most societies male
control of women's use of space is greatest during their reproductive
years, thus limiting their access to the labour market. Figure 8.3
shows that, although male economic activity rates vary little from
region to region, female rates show distinctive spatial patterns. In
China, although the female activity rate is only slightly less than that
of men, the decline with age starts earlier than elsewhere at about the
age of 35. Consequently Chinese women have the highest female

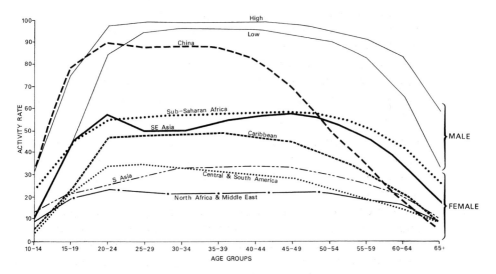

Figure 8.3 *Female and male economic activity rates over the life course, 1980.*

Source: based on International Labour Organization (1983) *Yearbook of Labour Statistics*, Geneva: ILO.

economic activity rate in the developing world between the ages of 10 and 50, but by the age of 65 the rate is lower than in any other region (Figure 8.3).

In most parts of the developing world women reach their maximum level of economic activity in their early twenties, while the maximum for men occurs a little later. In South-east Asia there is a marked dip in the level of female employment between the ages of 20 and 45, which is the period during which women experience the most intensive childbirth and child-rearing time demands. Other world regions do not demonstrate this so clearly. Its importance in South-east Asia may be a function of the high level of employment of

Box 8.1

Changing patterns of gendered economic activity in Brazil

Throughout the world more women are moving into paid employment as traditional restrictions break down, demand for new consumer goods spreads and economic pressure on families worsens. Brazil, the fifth largest country in the world, offers an example of the racial, regional and sectoral impact of globalization on gendered employment. Officially recorded female economic activity rates are usually higher in cities than in rural areas, especially in Latin America. In Brazil in 1940 only 18.9 per cent of the workforce was female and almost half worked in agriculture (IBGE 1990). The proportion of women in the workforce fell both relatively and absolutely in 1950 to 14.9 per cent but gradually recovered to 20.9 per cent in 1970, when half worked in the service sector and only 20 per cent in agriculture, reflecting both modernization of agriculture and urbanization (ibid.). During the 1980s women's participation in the labour force increased to 36.9 per cent in response to the economic crisis of the period (Calio 1990). In 1990 only five per cent of women workers were employed in agriculture compared with 18 per cent of male workers (IBGE 1994). The economic activity rate of women in urban areas was 40.1 per cent and 36 per cent in rural areas in 1990 (ibid.) and 41.3 and 39.4 respectively in 2000 (ECLAC 2002b). Women continued to increase their economic activity rate, predominantly in the service sector, especially in personal services and in education (IBGE 1990). Women provided 65 per cent of the workers in the personal service sector (domestic workers, cleaners, etc.) and 73 per cent of the social sector (teachers and health workers). Men workers were less concentrated by sector but both were moving into manufacturing, which occupied 23 per cent of men and 15 per cent of women in 1990. However, by 1999 industry had declined as an employer for both men and women, providing jobs for only 15.5 per cent of men and 9 per cent of women nationally. The gender difference in industrial employment was even more marked in metropolitan areas, with 19.3 per cent of men and 11 per cent of women in 1999 (IBGE 2002).

Women industrial workers, previously concentrated in the low-paid textile industry, began to move into the more modern metal and pharmaceutical industries. However, by 2000 the proportion of women in industry had declined to under 10 per cent, with 70 per cent concentrated in the service sector (World Bank 2001). Brazil has the highest proportion of women in professional and technical positions in Latin America: 61 per cent in urban areas and 76 per cent in rural areas in 1999, with only Uruguay having a similar employment pattern (ECLAC 2002b). Increases in educational opportunities enabled more women to move into professional positions and also reduced child workers' (ages 10 to 13) economic activity rates from 22 to 20 per cent for boys and from 10 to 8 per cent for girls between 1980 and 1990, despite economic pressures on households. This decline in child labour was balanced by an increase in economic activity by both men and women over 60, suggesting a family decision to invest in education. However, women household heads received only 55 per cent of the average wages of male household heads in 1991, rising to 62 per cent in 1995 (IBGE 2002), with the differential being greatest in agriculture despite little difference in hours worked.

Unemployment rates for women, higher than men's in 1981, had fallen below those of men by 1991 (IBGE 1994) but by the second half of the 1990s once again exceeded the rates for men (IBGE 2002). The highest-paid group in 1990 included 3.3 per cent of men workers but only 0.7 per cent of women workers, although 14 per cent of women but only 13 per cent of men had tertiary education (IBGE 1994). However, black women workers are the worst paid and the most likely to be in jobs without benefits (Calio 1990). Regional differences were also apparent, with women earning the highest wages compared to those of men in the frontier regions of the north and centre-west (85 per cent in Acre in 1995), but the highest proportion of unpaid women workers (59 per cent) was in the poor, largely Afro-Brazilian north-eastern region (Momsen 1992b; IBGE 1994, 2002).

Sources: Calio (1990); ECLAC (2002b); IBGE (1990, 1994, 2002); Momsen (1992b); World Bank (2001).

young, single women in factories, from which jobs they are fired when they marry or become pregnant.

In South Asia the maximum economic activity rate for women comes much later than elsewhere at age 45–9 and retirement also comes later, reflecting the early age of marriage and childbearing in this region. Early retirement from paid employment for women may occur because they no longer need the income as their children are grown-up and can support their parents. Or women may leave the labour force in order to take over the care of grandchildren and so release their own daughters for work outside the home. In any case these older women do not generally retire into idleness but take on new childcare and household responsibilities and often increase their labour input to the family farm.

The expansion of educational opportunities for women in recent years is reflected in the type of employment undertaken by women of different ages. Especially in Latin America and the Caribbean, where there has been a marked increase in women's access to education, younger women have moved into white-collar, urban managerial and administrative jobs, which offer regular employment, pensions and status. Their mothers generally continue to work intermittently in unskilled work, such as agricultural labour or trading. The greater financial independence of young women enables them to be less dependent on men and also less likely to see having children predominantly in terms of ensuring a future financial resource (see Box 8.1).

Industry

In most developing countries women have been moving out of agriculture and into industry faster than men. As a result the proportion of women in industry rose from 21 per cent in 1960 to 26.5 per cent in 1980, but by the late 1990s it had fallen, with more women moving into the services sector (World Bank 2001). On the other hand, in a few countries where export-oriented industries have recently expanded, the proportion of women employed in industry increased during the 1990s (ibid.). In many of these countries, such as Morocco, Honduras, El Salvador, Mauritius and Lithuania, there were more women than men working in industry by the late 1990s (UNDP 2002).

The influence of the international economy, as articulated by transnational manufacturing companies, has created a new market for female labour (see Plate 8.1). Manufactured exports from developing countries have become dominated by the kinds of goods produced by women workers. Industrialization in the postwar period has been as much female-led as export-led (Pearson 1998). However, the international economy has put a premium on low wages so the benefits to women of increased employment opportunities are equivocal. Some four million young women are employed in export-oriented industries in over 50 countries, mainly in South and South-east Asia and Latin America. This figure may underestimate the total as it does not include those who work informally for these transnational firms through subcontracting, piece-work and home-based work.

Plate 8.1 *China: young women workers making electric rice cookers for export in a factory near Guangzhou in the Pearl River Delta. This region, just inland of Hong Kong, had 30 million people working in manufacturing in 2003 who were paid on average 5 per cent of the mean American wage. The area attracts one billion dollars' worth of foreign direct investment and exports ten billion dollars' worth of goods each month. It has become the world's leading producer of items as varied as artificial Christmas trees, photocopiers and some of the best-known global brands of shoes. The main attraction is the cheap, mainly female, labour force, which is well disciplined and hard-working (Roberts and Kynge 2003).*
Source: author

The restructuring of the global economy associated with the new international division of labour has marked effects at global, national and local levels. Linked with this new spatial distribution of production is a restructuring of social relations, including gender relations, as labour markets recruit specific gender, age, ethnic and religious groups. These changes transform households, communities and markets and the changes in gender relations reflect shifting gender identities (Box 8.2). Parent–child relationships change as young women become the major earners in the family, and working in factories for transnational companies while living in urban dormitories with other young workers introduces rural women to new ideas (Wolf 1992).

Box 8.2

Negative perceptions of factory work for Muslim women in Malaysia

The most spectacular aspect of the growth of Malaysian export-oriented manufacturing was the massive and sudden involvement of young, single Muslim Malay women from rural areas. Between 1957 and 1976 the proportion of Malay women in the manufacturing sector trebled, exceeding that of the formerly predominant Chinese women. The proportion of Malay women working in factories rose from 19 per cent in 1975 to 26 per cent in 1979 and the proportion of men workers declined. This change was encouraged by the implementation of an ethnic quota under the Malaysian New Economic Policy after 1970 and created a backlash from both men and Chinese residents in factory areas. Both groups saw the young Malay women workers as a threat to the established social order. Some men saw the participation of women in industry as personal emasculation and expressed their anger by accusing the women workers of immoral behaviour. It has been suggested that similar behaviour by women employed in jobs recognized as 'women's work', such as typing or clerical work is ignored, while men feel that the foreign factories willing to employ women rather than men are a threat and so factory women are an acceptable target for public criticism.

Never before had Malay women left their traditional village occupations in such numbers. Most of these women came from families twice as large as the national average but with very low incomes. Three-quarters of the women chose to migrate to work in factories in order to reduce economic dependency on their households. Although factory wages were as low as those paid for agricultural work, they were more stable and offered fringe benefits, such as subsidized meals, medical services, transport to work, uniforms, sports facilities and other leisure activities. It was found that 56 per cent of the women migrated from their villages because they wanted to get a job and improve their standard of living, while a further 19 per cent did so in order to gain personal freedom and independence.

For the manufacturers these employees have many attractions. They are aged between 16 and 20 and so are more easily disciplined than older women. They are single and are thought to be more dependable than married women and more available for overtime assignments. They are poorly educated but not illiterate and the traditional rural compliance to male authority makes them the naive, obedient and malleable workers the firms want. Women working in manufacturing were paid only 69 per cent of male wages in the 1980s and this had fallen to 58 per cent by 1997 (Elson 2000: 93). They work 50 per cent more hours than women doing similar work in the West and receive only 10–12 per cent of the pay of Western workers. Low incomes lead to poor living conditions with overcrowding and few amenities. The combination of Western attitudes inculcated through factory work and living away from the protection of their family has led to involvement in social activities which are in conflict with traditional Malay Muslim values. Many parents are beginning to feel ashamed that their daughters are employed in an occupation that is rapidly acquiring a low moral and social status in Malaysian society.

In villages close to urban areas women can commute daily to factories, and in these settlements both family and community conflicts arise. In a study of mostly household heads in 45 villages in north-west Malaysia, 41 per cent had perceived in factory workers negative personal changes, such as indecent dressing, liberality in social mixing, decreased standards of morality, devaluation of domestic roles and loss of interest in local affairs. On the other hand, 12 per cent thought that the factory workers gained by increased knowledge and social exposure and being able to be self-supporting. Some 37 per cent of people interviewed thought that Malay women should be encouraged to work in factories, while 30 per cent felt the opposite. In relation to their own family members 11 per cent of these people were supportive of factory work, while 9 per cent opposed it. Overall, the source community accepted the utility of factory work in the short-term as an answer to immediate economic problems but rejected it in the long term because of the social and moral disutilities that were developing.

The moral stigmatization of Malay female factory workers is likely to continue as long as Malay Muslim society is not convinced that foreign-owned factories comply with Islamic standards of decency. In the study 79 per cent blamed the multinational value system, although 66 per cent felt that the women themselves had to accept some responsibility. Instead of merely ceasing to persuade Muslim women workers to participate in Western-style cultural and social activities, such as beauty queen contests, the companies would be more convincing to the local Muslim society if they embarked on activities to promote and safeguard local values. Changes introduced range from designing Islamically decent uniforms to providing prayer rooms in factories. Today many factories allow Muslim workers to take an additional 20 minutes during lunch and tea breaks to perform daily prayers. These workers have managed to persuade management that workers who enjoy inner peace and well-being tend to deliver higher productivity and quality output. However, factory women who choose to continue to participate in such alien activities as fashion pageants, annual balls and mixed picnics, which reinforce a feminine false consciousness, are now more vulnerable to criticism and negative perceptions, since the blame can no longer be placed on foreign factory managers. Furthermore, Muslim Malay society is not convinced that this new morality of the multinational companies is a genuine matter of faith and conviction and not merely an artificial device to pacify the women's parents so that their daughters are able to continue working in the factories.

Source: adapted from Buang (1993) and, updated by the author by e-mail, 20 January 2003; Elson (2000).

Plate 8.2   Thailand: young women weaving silk commercially in Chiang Mai.

Source: author

Plate 8.3   Thailand: older woman backstrap weaver in a village near Chiang Mai. Note the pipe-smoking.

Source: author

Women workers are concentrated in light industries producing consumer goods, ranging from food processing, textiles and garments to chemicals, rubber, plastics and electronics. In Egypt, Hong Kong, India, Kenya, the Philippines and South Korea over three-quarters of the female industrial labour force is employed in these seven industries. However, as manufacturing processes become more complex, men are increasingly being employed in manufacturing, especially in supervisory positions. In Bangladesh the proportion of women workers in manufacturing, in relation to male workers, fell from 328 per cent to 110 per cent between 1980 and 1990 but in most countries of Asia and the Pacific the numerical dominance of women in this sector continued to grow (ESCAP 1999). The growth of large-scale commercial manufacturing has resulted in older women's traditional craft skills becoming devalued (Plates 8.2 and 8.3).

Women also work in manufacturing outside the formal economy of the factories. Studies in Mexico City have shown that the number of women working in their homes, producing items on contract for factories, increased during the 1980s (Beneria and Roldan 1987). Women are employed to do simple, unskilled, labour-intensive tasks of assembly or finishing, requiring minimum use of capital or production tools. Working in the home allows women to carry out their productive and reproductive chores in the same location. The advantages of outworking for employers are the flexibility it gives them to respond to changes in demand and the reduction in labour costs. This work is on the edge of legality because of the absence of regulation, which enables employers to pay below minimum wage rates and to avoid providing fringe benefits and workplace facilities required by law. The work offers no security but may be the only or best option for women trapped in the home with young children.

Women also work as petty commodity producers in both rural and urban areas. Like outworking, self-employment offers women flexibility of time and space as it can be combined with domestic chores (Plates 8.4, 8.5 and 8.6). In traditional societies it may be more acceptable for women than working for someone else outside the home. Sometimes production may be done within a community, where women work in a shared space but retain individual rights to their production. In a Sri Lankan village women of Tamil, Singhalese and Moor ethnicity and Hindu, Buddhist and Muslim religious affiliations meet on one woman's verandah to make local cigarettes and paper bags together, thus providing cross-cultural social

Plate 8.4
*Burkina Faso: women
potters in a north-western
village. These pots are
used for carrying water.
The thatched buildings in
the background are
granaries for storing grain,
the square flat-topped
building in the centre is
for storing pots and the
smaller building in front of
that is a hen house.*

Source: Vincent Dao, University of
California, Davis

exchange and sharing responsibility for childcare while individually earning (Ismail 1999a). The types of goods produced by women generally vary from textiles and garments involving weaving, lace-making, sewing and embroidery, to ceramic and food items (Plate 8.7). Self-employment builds on women's traditional skills and has been expanding recently as aid organizations offer assistance in the form of credit, training, design and marketing.

Microfinance

Provision of microcredit has been seen as the way to help women to set up small businesses and to be empowered. The Grameen Bank in Bangladesh led the way in 1976 in providing small loans to poor

Plate 8.5
Guatemala: spinning cotton.
Source: author

women. Repayment rates were very high, largely because the women had to borrow as a member of a group and other members could not get new loans until old ones had been repaid. By 2000 the Grameen Bank was lending money to 2.37 million borrowers in over 40,000 Bangladeshi villages at an interest rate of between 20 and 35 per cent (Akhter 2000). The average loan size is US$160 and the repayment rate is 95 per cent (ibid.). The Grameen model has been replicated by 223 organizations in 58 countries. The Microcredit Summit +5, held in 2002, saw microfinance as being pro-poor and set a goal to reach 100 million of the world's poorest families with credit for self-employment and other financial services by the year 2005.

At first this system was seen as a very positive contribution to development, especially for poor rural women. Microfinance was

Plate 8.6 *Peru: basket making. Note the multitasking of childcare and handicraft production.*

Source: Rebecca Torres, University of East Carolina

promoted as a self-help 'human face' complement to structural adjustment in the context of declining aid budgets and reduction in government subsidies (Mayoux 2002). However, the incomes earned from small-scale self-employment, such as is supported by microloans, are rarely sufficient to pay the increased costs of basic consumption goods and services (ibid.). The World Bank has argued that Grameen Bank loans to women tend to lead to an increase in girls' schooling and in per capita consumption, a reduction in fertility and increases in women's paid work and nonland assets (World Bank 2001). Other studies of the impact of loans on recipients have revealed problems such as the women's loans being used by husbands and the feminization of indebtedness (Akhter 2000). Having a loan does not always empower women nor give them decision-making power within the family and the impact often depends on the skills of the individual woman (Kabeer 1998). Loans are most successful when they include training. Furthermore, by perceiving women's economic activities as only occurring within the informal or small-scale sector, it might be suggested that they are being forced into a microcredit ghetto (Randriamaro 2001).

Plate 8.7 *Brazil: lacemaking in a coastal village in Ceará. The woman is working, while her husband, a fisherman, is at sea, in order to help support the family. Most of her production is sent to São Paulo, the biggest city in Brazil, where it fetches high prices.*

Source: author

It appears that the Grameen Bank has benefited more than women from the gender bias in the loans, as women are seen as being more reliable and responsible and more susceptible to peer pressure than men, so making them better at repaying loans. The Grameen Bank has also used these loans to link villages with the global economy, particularly through their programme of providing mobile phones to villagers in areas with no fixed phone lines and no other cellphone providers to offer competition. Some of these for-profit programmes are more questionable, such as the linkage with Monsanto, since ended, which forced loan recipients to buy GM seeds and hybrid chicks, which often do not do well in rural Bangladesh (Akhter 2000).

Female targeting may be a very cost-effective strategy for microfinance institutions. It may also increase the well-being of children, though it excludes the poorest 5 per cent of the population who are considered too poor to receive loans. However, there are serious dangers that female targeting without adequate support networks and empowerment strategies will merely shift all the burden of household debt and subsistence and even of development itself on to women (Mayoux 2002). Questions are being raised about the 'Grameen' system of microfinance as it has spread throughout the South. Are women being used by men in households in order to get access to credit for men and to allow men to reduce their own contributions to family expenses? Are men encouraging women to take the lead in microfinance because men do not want to put in the time and effort to attend meetings? Are programmes targeted at women because they are a more docile clientele whom it is easier to pressure into repayment? Are the main beneficiaries the programme staff who can be paid higher salaries and the institutions that can build new offices in urban centres from the high interest rates paid by poor, often rural, women? Is self-help for poor women in the South promoted by northern governments to avoid addressing inequalities in trade and aid relations, and by southern governments to avoid dealing with wealth redistribution and legal and political reform (ibid.)?

Recent fieldwork in central Sri Lanka has shown that, although microcredit has only a minimal impact on poverty, it does empower women by enabling them to earn money that is not controlled by their husbands and to go out alone to group meetings, forcing men to take over some domestic tasks (Aladuwaka 2002) (Plates 8.8 and 8.9 and 8.10). It also encourages women to participate in community management and empowers them to work with other women to reduce male alcoholism in the community (ibid.). The successful entrepreneurs in this community felt that they gained new respect from their husbands by earning their own money.

New issues have arisen as the microfinance system has spread from its origins in Bangladesh. In Latin America, poor women trying to work in the informal economy are more likely to be found in urban slums than in rural areas so microfinance has become more urban focused. Non-profit organizations showed banks that poor people are good at paying back their loans, so private banks have stepped into the market, moving from underwriting 10 per cent of loans in 1995 to 60 per cent in 2000 (Chacón 2000). In Africa and the Caribbean there is a long tradition of informal savings, credit and insurance arrangements, such as Rotating Savings and Credit Associations

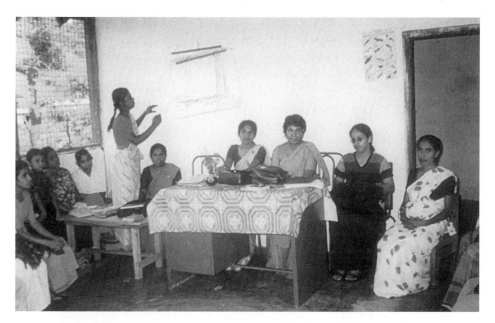

Plate 8.8 *Sri Lanka: Samurdhi microcredit group meeting in a village in the Kandy area.*
Source: Seela Aladuwaka, University of Peradinya

Plate 8.9 *Sri Lanka: a woman dairy farmer. She used her loan to buy the cows to start her business and is making a profit from selling the milk.*
Source: Seela Aladuwaka, University of Peradinya

Plate 8.10 *Sri Lanka: this woman started her shop with a loan. It is her first independent economic activity and gives her and her family some income. Her husband works with her in running the shop. She combines her business with her family responsibilities but feels that the business has given her more respect in the family. She plans to take another loan to expand her store.*

Source: Seela Aladuwaka, University of Peradinya

(ROSCAs), and women are used to working with other women in groups. Microfinance may offer more confidentiality and security to build up independent savings than ROSCAs, and in Asia and Latin America the savings groups often provide socially acceptable places for women to meet. In Africa there is less control of women's mobility and in many African countries women have extensive networks through ROSCAs, church groups and market associations, and may not have the time to give to savings groups. New groups set up by outside funding agencies may destroy pre-existing groups, as they did women's fishtrading associations in southern Ghana, because the agency-sponsored groups are exclusive and inflexible (Walker 1998). In Bangladesh Goetz and Gupta (1996) have shown that the embeddedness of local patronage networks makes it difficult for fieldworkers to assign group membership to the most needy. In addition, employees of funding agencies are often suspected of

Christian evangelizing, breaking up families or involvement in the trafficking of women (ibid.).

A study in India and Bangladesh (Hunt and Kasynathan 2001) found that NGO staff assumed that access to credit automatically led to an improvement in women's status in the household but had few hard facts to support this assumption. It was found that NGO staff tended to overestimate the amount of control women have over their loans and there was no mechanism to monitor the degree of control, or the impact of microfinance on violence against women, dowry, divorce and polygamy. Women are more likely to control their loans if their husbands are absent or if the money is used for a 'traditional female activity'. There is some evidence that access to credit may lead to more schooling for daughters but this has not been studied in detail and may be related to other changes in rural communities, such as greater accessibility of schools. Credit may not increase women's mobility as much as expected and other factors, such as extreme poverty, may be more influential. The study suggests that NGOs need to undertake closer monitoring of control of credit and to provide technical training to the recipients of the loans in financial management and marketing (Hunt and Kasynathan 2001). Overall, there is a tendency to assume that all positive aspects of rural development are related to microfinance.

As the use of microfinance has spread worldwide, the emphasis on gender empowerment has been overtaken by a focus on the financial sustainability of lending institutions. To make microloans more effective they need to include family loans, where both husband and wife are responsible for repayment, to incorporate training to improve effective credit utilization, and to allow for flexibility in repayment schedules to take into account seasonality in economic activity patterns (Mayoux 2002). Microfinance should be seen as a component of but not a substitute for a coherent agenda for poverty elimination.

Gendered employment in the service sector

Estimation of the gender pattern of work is particularly difficult in this sector. Even so, official statistics indicate that women make up 68 per cent of the service workers for the upper middle-income countries and 73 per cent in Latin America and the Caribbean (World Bank 2001). In general women work in health, education, catering, tourism and commerce at the lowest and worst-paid levels. Men are more likely to be working in professional and transport services.

Informal sector work

Providing services in the informal sector involves many women. Work in this sector as traders, servants or prostitutes is often the only urban employment open to young, uneducated women from rural areas. There are distinct regional patterns in the dominant types of employment.

Women are especially important in retail trade in Africa and the Caribbean: they make up 93 per cent of market traders in Accra (Ghana), 87 per cent in Lagos (Nigeria), 60 per cent in Dakar (Senegal) and 77 per cent in Haiti. Status as a market trader comes with maturity and mothers often pass on to daughters their bargaining skills and the goodwill of their customers. Although most traders in West Africa are women, they tend to concentrate on the sale of small quantities of home-produced items in local markets, while men control wholesaling and the long-distance trade in manufactured goods. Similar distinctive gender roles can be seen in the Caribbean, where beach vendors selling to tourists fall into two groups: young men who sell jewellery or suntan oil and work as vendors for a few years because they enjoy the beach and the chance to meet young female tourists; and older women who braid hair or sell home-made clothing on the beach because the job does not have fixed workhours and can be combined with childcare.

In Latin America, domestic service occupies many women. It is seen as an entry point into urban employment for female migrants from rural areas (Momsen 1999). Their lowly occupational status may be reinforced by racial discrimination. In Andean and Central American countries, servants are often Indian and may speak little Spanish, in contrast to the Hispanicized families for whom they work. Domestic service is distinctive in that women are both employer and employee. Employers prefer young rural women whom they can train and often develop a complex relationship with their servants based on both dependency and exploitation (Tam 1999) (see Box 7.2). Middle-class, professional working women in rich countries hand over their burden of housework and childcare to lower-class women but then are dependent on these servants, many of whom are international migrants (Momsen 1999; Mattingly 2001). Where these migrants do

Box 8.3

New aspects of prostitution in Thailand

Men are now moving into prostitution in Bangkok, with numbers tripling since 2000, reaching 30,000 in 2002. Male sex workers are mainly confined to the tourist areas of Bangkok, Chiang Mai and Phuket. Some are as young as 12 and have gay men visiting from abroad as customers, while those catering to foreign and Thai women are older. Clients include locals and foreigners, homosexual men, transsexuals, bisexuals and women. It has been suggested that the number of women clients is increasing, which reflects a growing attitude among Thai women that they should get even with their philandering husbands. Male prostitutes have lower HIV/AIDS prevalence rates than female sex workers: official figures show that 9.6 per cent of male sex workers are HIV positive compared with 16.2 per cent of female prostitutes.

The Thai government rewrote its prostitution legislation in 1996 in response to international pressure to stamp out the sexual exploitation of children. For the first time the law made criminals of the clients rather than their young victims. Now the regime at the government-sponsored children's home has changed and the centre promotes itself as a provider of physical, psychological and emotional rehabilitation. The girls are taught new skills which may equip them for legal jobs so that they are not forced to return to the sex industry when they leave. They are taught traditional skills, such as playing Thai musical instruments and hand weaving, as well as high-tech skills, such as desktop publishing. But for every child that is rescued many more are still suffering. Thailand's National Commission on Women's Affairs estimates that 30,000 to 35,000 children, mostly girls, work in the country's sex industry. This is probably an underestimation and some observers think the figure could be ten times that number.

Despite the new law, action against clients has been very slow. It is suspected that law enforcement authorities are linked with traffickers in some areas and so are turning a blind eye to the trade. Thailand continues to act as a magnet for young people from poor rural families in neighbouring countries, such as Laos, Myanmar and Cambodia, as its average per capita income is much higher than in those poor countries. A prostitute can earn ten times the daily income of a factory worker, more if she/he agrees not to use a condom. Thus individuals may choose to go into the oldest profession for economic reasons, and are not always tricked by local agents or forced by traffickers or criminals to whom they have been sold by poor rural parents. However, the Thai government has embarked on an ambitious programme to change male attitudes towards women. This has apparently diminished the number of visits to local brothels in a surprisingly short time (UNAIDS 2000).

Sources: adapted from Denny (2000) and UNAIDS (2000).

not have legal status they are open to even greater exploitation (Anderson 2000). Thus the two circuits of formal and informal employment and of domestic and international migration are intertwined, especially for women.

Working in private households, maids are unlikely to be protected by employment legislation, may be expected to work very long hours and may also be exposed to the sexual advances of the male members of the household. If they become pregnant they will lose their job and may have to turn to prostitution. Sometimes they will return to their natal rural villages to get married or may stay in the cities and find better-paid jobs. Foreign maids will be forced to return to their own countries or to work illegally.

In South-east Asia the provision of sexual services employs many women, especially in areas with large foreign military bases and a tradition of men seeking sexual gratification outside marriage. However, there is some evidence of a backlash from wives in that they are beginning to utilize male prostitutes (Box 8.3). Prostitution is becoming more diverse, involving men and women, boys and girls, and homosexual and heterosexual sex.

In African cities prostitution is generally less organized than in Asia and many of the women work on their own account. This may enable women to keep more of their earnings but also means they are offered less protection. AIDS is a greater problem in Africa than in Asia and was already spreading quickly among prostitutes in Nairobi, Kenya, in the 1980s. In many Asian countries special attention is being paid to HIV/AIDS infection among this group of workers by encouraging the use of condoms and providing regular health check-ups. Training, so that prostitutes can move into other occupations, is also being offered to a limited degree. Unfortunately, most of these alternatives, such as craftwork or taxi driving, are not as profitable as prostitution. Growing fear of disease is leading to a premium being charged by brothel owners for virgin and child sex workers.

Some countries, such as the Philippines and Thailand, have developed a tourist industry which exploits the trade in female sexuality. The sex trade is one of the few ways in which foreign women can live in Japan. Sri Lanka has developed a reputation as a source of child sex and so attracts foreign paedophiles. Today many countries do not enforce laws against prostitution because of its importance to tourism. However, some Brazilian beach resorts are discouraging single male tourists because of the link with paedophilia. In addition, a few European countries are beginning to prosecute their own nationals who are caught overseas in what would be considered illegal sex acts in Europe.

Tourism

Improvements in global communications and transport and in infrastructure in poor countries has encouraged a period of explosive grown in tourism, especially in developing countries. Tourism is seen as a new export-led growth industry and many countries of the South have very rapidly become dependent on tourism. This is especially true for small island states with few natural resources except sun, sand and sea, as can be seen in the Caribbean and the South Pacific. Employment indirectly stimulated by tourism, such as the production of craft souvenirs, may also benefit poor communities by encouraging the survival of traditional crafts because of increased demand (Plate 8.11). It has been argued that these traditional crafts provide the authenticity sought by tourists, although Western ideas of design and the photogenic may lead to changes in the products so that they are no longer truly authentic (Swain and Momsen 2002). Production of souvenirs is flexible work in the informal economy which is often undertaken by women. However, greater financial returns for craftwork may encourage change in gender roles, with men becoming dominant in production as in Peru or taking over the more lucrative marketing of crafts as in Malta or Indonesia (ibid.). On the other hand, women may be able to exploit their photogenic appearance to obtain economic benefits from visitors.

Women also work in the formal tourism sector, making up an average of 46 per cent of the world tourism workforce (Michael et al. 1999). The proportion of women tourism workers varies widely, from 77 per cent in Bolivia and 75 per cent in Peru, Botswana and Estonia to under 5 per cent in some Muslim countries (ibid.). Tourism provides mainly low-paying jobs for women as maids and housekeepers or as receptionists in hotels, while the more lucrative positions of managers and chefs are predominantly male. Women often take on these low-skilled positions because these are the only positions available to them in many isolated rural or specialist tourist areas. Also, despite the low pay, such jobs offer the flexibility needed by women with children and may allow families to continue living in a location to which they are attached, in the

Plate 8.11 *French Polynesia: women preparing flower garlands for use in welcoming tourists. They are sitting outside the main tourist attraction, the Gauguin Museum, in Hiva Oa, Marquesas. This occupation is combined with childcare.*

Source: author

pleasant environment which has attracted tourists. Often the only alternative jobs available in the area are in agriculture, where the pay is as low and the work even more seasonal and exhausting. In addition, menial hotel work, where there is a modern, attractive, air-conditioned environment, a smart uniform and the possibility of tips and small gifts and the opportunity of meeting people from many different places, may be more attractive and pleasurable than work in field or factory.

Tourism is supposed to be undertaken for pleasure, relaxation and the enjoyment of new experiences. It is separated from everyday life and thus provides a refreshing break from the pressures of daily living. Yet this separation does not mean that gender differences disappear. All social experience is gendered and tourism as an industry and activity influences gender roles and informs gender relations (Kinnaird and Hall 1994).

Tourism can be a gendered social catalyst. The process of bringing hosts and guests together in a single location introduces a cross-cultural exposure which may induce changes in perceptions and behaviours of individuals. This can especially be seen in rural tourism, which is growing rapidly in many transition countries. Foreign visitors to guest houses and farms usually have most contact with women hosts who provide meals and guidance to the local area. Such interaction may be the basis of long-term friendships and provides very welcome social contact for women in isolated areas.

The more different the host and guest communities, the longer the stay, and the more rapid the growth of tourism, the greater the impact in general. These cultural border contact zones or global margins within which the interactions between tourist and local take place form new spaces of modernization and change. They may be exploited to mutual benefit but, in some cases, contact may cause merely confusion and misunderstanding on both sides. Individual responses to tourism may lead to changes in gender roles and in domestic relations within the household. Increased social capital, in terms of wider networks external to family and community, may be a major benefit to women as they gain at least some economic independence through employment in tourism.

Tourism involves, more than most industries, face-to-face interaction. Holiday pleasure often depends on the caring and hospitable nature of the hosts. Thus women's employment in tourism is not just the result of the willingness of women to work in low-paying jobs which utilize their supposedly 'natural' housekeeping skills, but also of their ability to provide friendly care and assistance to guests. Thus the 'managed heart' identified by Hochschild (1983) is the stereotype which is seen as making women particularly suitable for running guest houses, or for working as recreation directors in big hotels and even as waitresses (Plate 8.12). Sexual objectification of women is found in many aspects of the industry, with female hotel employees in particular being told how to dress and comport themselves in much greater detail than is true for male employees. Women in the hospitality industry may also be expected to flirt with guests in order to encourage additional consumption at the bar, for example. Thus sexuality can constitute an element of gendered economic relations.

The imposition of Western consumption patterns on traditional societies causes conflict. Governments in countries of the South, faced with falling prices for commodity exports and a balance of payments crisis, may encourage patriarchal family control of women in order to offer foreign companies and male tourists cheap, skilled

Plate 8.12
China: the 'managed heart'.
Minority woman offering
apples to foreign visitors in
the mountains of Yunnan
province.
Source: author

young women (see Boxes 8.2 and 8.3). Thus national prosperity has
been seen to depend on a continuation of female subordination and
poverty but there are signs that women in many places are rebelling
and taking control of their lives.

Learning outcomes

- The influence of globalization, articulated by transnational companies, has created a new market for female labour.
- Microfinance offers opportunities for women to develop entrepreneurial skills but may also be the cause of the feminization of debt.
- Tourism is the fastest-growing industry in the world and is increasingly dependent on the 'managed heart' of women.
- Women's employment patterns vary by age from country to country more than men's.

Discussion questions

1 Why do women predominate as workers in factories in export-processing zones?

2 Outline the advantages and disadvantages to women of work in the informal sector

3 Rehearse the reasons for differences in male and female economic activity rates at different ages.

4 Microcredit provides both opportunity and problems for women. Explain.

5 What is meant by identifying tourism as a gendered social catalyst?

Further reading

Afshar, Haleh and Stephanie Barrientos (eds) (1999) *Women, Globalization and Fragmentation in the Developing World*, London and Basingstoke: Macmillan; New York: St Martin's Press. Provides a range of case studies of the impact of globalization on women's employment in developing countries.

Anderson, Bridget (2000) *Doing the Dirty Work? The Global Politics of Domestic Labour*, London: Zed Books. Looks at the problems faced by women migrants from poor countries working as domestics in developed countries.

Beneria, Lourdes and Martha Roldan (1987) *The Crossroads of Class and Gender: Industrial Homework, Subcontracting and Household Dynamics in Mexico City*, Chicago: University of Chicago Press. A classic study of women in the informal economy.

Kinnaird, Vivian and Derek Hall (eds) (1994) *Tourism: A Gender Analysis*, New York and Chichester: John Wiley and Sons. The first edited collection on gender and tourism, with examples from South-east Asia, the Caribbean and Western Samoa as well as Europe.

Swain, M. B. and J. H. Momsen (eds) (2002) *Gender/Tourism/Fun(?)*, New York: Cognizant Communication Corporation. A collection looking at the gendered enjoyment of tourism by both tourists and workers in the industry.

Websites

www.oneworldaction.org   One World Action based in London has information on microcredit posted on its website.

www.world-tourism.org   The World Tourism Organization is an intergovernmental body charged by the United Nations with promoting and developing tourism.

www.globalisation.gov.uk   This site provides the UK's Department for International Development (DFID) White Paper on 'Eliminating world poverty – making globalisation work for the poor'.

9 How far have we come?

Learning objectives

At the end of this chapter you should understand:

- the extent of progress in gender equality
- gender roles in politics and the value of quotas
- the impact of poverty in terms of both material assets and individual well-being
- the new economy and its gendered impact.

As Boutros Boutros-Ghali said: 'No true social transformation can occur until every society learns to adopt new values, forging relationships between men and women based on equality, equal responsibility and mutual respect' (1996: 73). Although there is formal recognition of women's rights and legal equality, the gender disaggregated data that are now increasingly available have revealed that women continue to face discrimination (Fenster 1999). Most governments have formally adopted the 1979 Convention on the Elimination of All Forms of Discrimination against Women (CEDAW) but the implementation of the CEDAW principles is far from complete. It is also worth noting that the number of reservations expressed by governments adopting the Convention was the highest for any human rights instrument negotiated under the auspices of the United Nations, indicating the obstacles still faced by women (ibid.: 72). In Africa, many legal rulings concerning women's rights continue to follow customary law rather than international

agreements or even constitutional provisions enshrined within the bill
of rights and the basic legal regime of a country, further undermining
CEDAW (Oloka-Onyango 2000).

With the ending of the cold war the attention of the international
community turned to a range of long-neglected global problems,
including the position of women. The 1990s saw an unprecedented
series of conferences, some directly concerned with women and
others on environment and social issues in which women's voices
played a major role in decision-making. Yet, at the national level,
the decision-making position of women deteriorated. In 2001 women
held 14 per cent of all seats in parliaments, an increase from 9 per
cent in 1987. But the proportion for the less developed countries was
only 12 per cent, compared to 18 per cent in the more developed
countries (PRB 2002). In 2002 the proportion varied from 42.7 per
cent in Sweden to zero in several Middle Eastern countries (UNDP
2002), although women in Bahrain did get both to vote and to stand
as candidates for the first time (Figure 9.1). Many of the political
inroads women have made are due to gender quotas designed to seat
more women in legislative bodies, from national parliaments to
village councils. A number of countries, including Brazil, France,
India, the Philippines and Uganda, have established such quotas in
the last decade (PRB 2002). In South Africa, following the end of
apartheid in 1994 and the adoption of a new constitution that
promotes women's rights, the proportion of women in Parliament
rose from 1 per cent to 30 per cent. On the other hand, in the former
communist nations of Eastern Europe, quotas for women were
abolished, resulting in a dramatic fall in the number of women
elected from 25 per cent to 7 per cent (World Bank 2001). It has
been suggested that, compared to economic opportunity, education
and legal rights, political representation is the aspect in which the
gender gap narrowed the least between 1995 and 2000 (Norris 2001).
In 1996 there were 24 countries that had had women or currently had
women as elected heads of government (Seager 1997). In seven
countries of the South in 1995 women held one-fifth or more of
cabinet-level appointments but there were 59 countries with no
women in cabinet positions and only two developing countries, the
Seychelles and Mozambique, with more than a quarter women
elected officials (ibid.). By 2000 there were 11, widely distributed,
less developed countries with over 25 per cent women
parliamentarians, of which six were in Africa, three in Latin America
and the Caribbean, one in South-east Asia and one in Central Asia

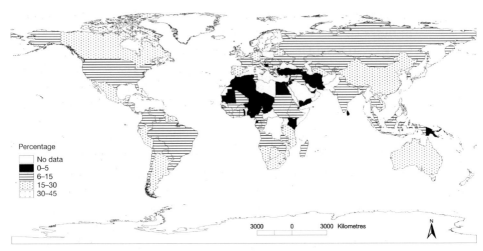

Figure 9.1 *Women as a proportion of national elected legislators, October, 2001.*
Source: Sass and Ashford (2002: 38–45)

(UNDP 2002), but 20 with more than a fifth of ministerial positions held by women. Interestingly, of these 20 countries only three (Grenada, South Africa and Uganda), with several women in policy-making positions, were the same countries as had large numbers of elected women (PRB 2002).

Increasing the political representation of women is often considered to be a way of improving the state's success in meeting women's needs and raising the efficiency of government as women politicians are considered to be less corrupt and more altruistic than their male counterparts. Thus the United Nations and the World Bank, in their publications, tend to see more equal gender representation in politics as contributing to development. Today over 30 countries from Morocco to East Timor have established quotas for women in national legislatures. However, high political position does not necessarily lead to anti-discriminatory legislation. Women elected on quotas in Eastern Europe had little power, and in countries such as Romania and the former Yugoslavia the most powerful women achieved their influence as wives of presidents and were probably more dangerous for women than their husbands (see Chapter 3). In South Asia, where discrimination against women is still very strong, Bangladesh, India, Pakistan and Sri Lanka have all had women political leaders but they have gained their positions based on their elite status and family connections rather than on their own personal qualities.

It has been suggested that women living through the post-communist transition question the value of seeking public roles because they do not see the state as a site of liberation. Rather, they see the private space of the family as the source of agency, since this offered them the most freedom under communist regimes (Marchand and Runyan 2000). At the local government level, in the transition countries and in South Asia, however, there is a trend towards more elected women. Thus change is slowly coming at the grass-roots level and its effect on improving women's lives and local development is noticeable. Women mayors in Hungary are more likely to focus on improving the environment and appearance of their towns and to encourage tourism, an occupation usually involving more women than men, than are male mayors (Szorenyi 2000). In India, political reservation of seats on village councils for women has influenced policy-making. On forest conservation committees women were at first too shy to speak out (Bode 1993) but Agarwal (1997b) has shown that eventually the rules established by these committees were different, where women had participated in their elaboration, taking into account women's needs to collect fuel. In a study of West Bengal, where the position of council leader on one-third of village councils has been reserved for women since 1998, Chattopadhyay and Duflo (2002) showed that, where the leader of the council was a woman, village women were more likely to participate in the policy-making process. Village councils with women leaders invested more in matters directly relevant to rural women, such as roads, fuel and water, and were more interested in overseeing the performance of village health clinics than of informal village schools. The greater interest of men council leaders in education may be a reflection of their generally higher educational level. It has been shown for Latin America (ECLAC 2002a) that uneducated mothers are less likely to push their children to stay in school and it may be that Indian women council leaders did not give the strategic utility of education as high a priority as more immediate practical needs that could reduce the burden of daily work for village women. Above all, women continue to be involved in political protest to protect their families and the environment or improve living conditions (Radcliffe and Westwood 1993; Rocheleau *et al.* 1996).

The state as a collection of institutions partly reflects and partly helps to create particular forms of gender relations and gender inequality. State practices construct and legitimate gender divisions, and gender identities are in part the result of legal restrictions and opportunities emanating from the state (Waylen 1996). Thus the state plays a key role in regulating gender relations. State policies, established by generally patriarchal institutions, are gendered according to their subject matter. Three types can be identified: those policies directed towards women, such as reproductive rights; those dealing with the balance of power in gender relations, such as marriage and property rights; and those ostensibly gender-neutral policies, which, however, affect men and women differently, such as resource extraction and social reproduction.

Gender and development planning

One of the themes running through this book is the gender-blind nature of much development planning and its failure to consider both women's and men's needs and viewpoints. Women are central to development. They control most of the non-money economy through bearing and raising children, and through providing much of the labour for household maintenance and subsistence agriculture. Women also make an important contribution to the money economy by their work in both the formal and informal sectors but these roles are often ignored. Everywhere in the world women have two jobs – in the home and outside it. Women's work is generally undervalued and the additional burden development imposes on women is usually unrecognized. Their health suffers, their children suffer and their paid work suffers. Development itself is held back. As poor countries face new problems women's roles become increasingly influential. In addition, development often brings greater flexibility to gender roles and changes gender norms.

The gender bias in development has often been interpreted as a failure to include women. Recognition of the importance of the gender gap led to the 1995 establishment of the Gender-Related Development Index (GDI) (Figure 9.2). This combines gender-related measures of life expectancy, adult literacy, enrolment in primary, secondary and tertiary education, and estimates of earned income, to arrive at a country-by-country evaluation of the gender gap in achievement. The GDI adjusts the Human Development Index downward, based on the belief that gender inequality reduces the overall level of well-being in a country (Bardhan and Klasen 1999). The extent of this downward adjustment is almost entirely determined by gender gaps in the earned-income component. Changes in the calculation of the income component in 1999 reduced

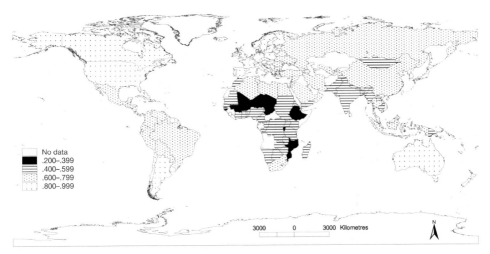

Figure 9.2   *The Gender-related Development Index, 2001.*

Source: UNDP (2001: 210–13)

the disparity between the HDI and the GDI (Bardhan and Klasen 2000), with Saudi Arabia and Oman having the greatest gender gap of –10 and Jamaica the highest positive difference of +6. The rich countries of the North rank highest on the GDI, with Slovenia at 27, followed by other transition countries, Caribbean states and Latin America, and, in the lowest ranks, the states of South Asia and Africa (UNDP 2002) (see Figure 9.2).

Changes in the GDI are hard to measure over time because of different methods of calculation used, but during the 1990s 21 countries registered a decline in their Human Development Index. Those countries with the lowest GDI rankings are largely very poor African countries which are currently or have been recently affected by civil war, with many women and children being forced to become refugees. It is noticeable that the transition countries of Eastern Europe, and Russia (ranked at 60) plus China (rank 96), where under communism there was official gender equality, now rank well below the richer countries. These countries have a similar rank to the predominantly Muslim countries of North Africa and the Middle East, where women now have access in many cases to education but little earned income. The transition countries rank higher on the GDI than on the Human Development Index (HDI), while the reverse is true for the Muslim countries. The biggest differences in rankings on these two indices are for Saudi Arabia and Oman, where national wealth is not reflected in gender equality

(UNDP 2002). Although there are clear links between the level of economic development and the gender-related index because of the importance of the income measure, poorer countries with gender equality of education, such as Jamaica and Sri Lanka, rank higher on the GDI than on the HDI.

However, women's contributions will never be equally valued until women's priorities are included as objectives of development (Kabeer 1994). This challenge highlights the importance of another index, the Gender Empowerment Index (GEM), which takes into account the political power of women and the proportion of women in professional and technical jobs. Barbados, at 18, is the highest-ranking developing country on this index, ahead of the transition countries. Unfortunately, data was only available to rank 66 countries on the GEM index compared to 146 on the GDI index at the beginning of the third millennium (UNDP 2002).

Many of the resolutions passed at the international women's conferences in Nairobi and Beijing are based on a paradox. They reflect the expectations that national governments are responsible for implementing these international commitments and introducing policies aimed at improving the lot of women. But they do not address the ways in which market liberalization and privatization may undermine the ability of governments to discharge these responsibilities (Elson 2000). This is especially true in poor countries faced by disasters, such as the spread of HIV/AIDS. Respecting the human rights of women and men and of minorities will help to empower individuals and reduce violence. But recent upheavals, such as civil wars and economic turmoil, have undermined human rights in many developing countries (Olcott 2000; Oloka-Onyango 2000; Fenster 1999; Silovic 1999).

Contemporary problems of development

Since the early 1980s many poor countries have found themselves with an increasing debt burden. This has been particularly the case in Latin America and the Caribbean and in Africa, while Asian countries had generally lower debt burdens, especially from private sources (Feldman 1992). Often the heavily indebted nations have been forced to ask for financial assistance from the International Monetary Fund (IMF). In return for this assistance the IMF usually imposes tough financial constraints, and demands that the recipient

Table 9.1 *Seasonal changes in women's time spent on different activities under conditions of environmental and economic stress, in a village in northern Ghana, 1984 and 1991*

Activity	Average hours per day			
	Wet season		Dry season	
	1984	1991	1984	1991
Reproductive work:				
Household maintenance*	5.6	4.6	6.4	5.1
Kitchen gardening	1.5	1.4	0.6	0.4
Subtotal	7.1	6.0	7.0	5.5
Social duties	0.9	0.5	2.4	1.7
Productive work	3.3	4.1	2.3	5.2
Average hours per day for those involved in productive work:				
Farming	3.5	5.2	1.1	1.6
Trading	3.7	4.8	1.0	6.3
Brewing	2.7	4.8	3.9	4.2
Food processing	4.8	4.7	3.0	5.8
Handicrafts	2.0	5.0	2.6	5.3
Formal sector	0.0	7.0	0.0	7.0
Total average hours/day	11.3	10.6	11.7	12.4

Note: * Household maintenance activities include food processing, preparation and storage, child and elder care, water and fuelwood collection, laundry and housework.

Source: Awumbila (1994: 275 and 278).

countries restructure their economies. This structural adjustment generally involves an increase in production for export combined with demand-reducing policies, such as removal of subsidies on basic foodstuffs, reduction in welfare services, higher charges for basic services, price rises, wage cuts and job losses. It may be argued that the social costs of structural adjustment would have been worse without the IMF intervention, but for a long time there was little appreciation by international agencies of the gender bias in their impact.

In response to structural adjustment, women have developed new survival strategies. This behaviour has been called 'invisible adjustment', implying that women make adjustment policies socially possible by increasing their own economic activity, by working harder and by self-abnegation. Table 9.1 shows the impact of the introduction of charges for use of the village water pump and for schooling in a village in Ghana in 1991. There was a marked increase in the time women spent on income-earning work at the expense of social duties, household maintenance and their own leisure, especially in the dry season.

The hegemony of neoliberal structural adjustment shifts the burden of welfare from the state to individual families and especially women (Afshar and Dennis 1992). Wage cuts and the rising cost of living force more members of the family to seek paid employment and, because women are paid less than men, it may be easier for them to find employment (Beneria and Feldman 1992). The impact of the economic crisis of the 1980s on female economic activity rates was very mixed. It varied not only from country to country but within countries, between economic sectors, between urban and rural areas and according to age and educational levels. Increased production for export provided new jobs for women in labour-intensive manufacturing, data entry and word processing. Some new jobs were also created for women in agribusiness, but the expansion of agricultural exports benefited men more than women (Joekes and Weston 1994). The most widespread effect was a slowdown in the

Plate 9.1 *East Timor: pounding corn in old shell cases.*
Source: author

Plate 9.2 *Sri Lanka: young daughters helping with household chores. Hanging out washing in a rural area near Kandy, while parents are working in the fields.*
Source: author

participation of women in the formal workforce, which had been growing since 1960, and an expansion of the informal economy. Family structure also responded, with an increase in the number of extended families in which both housework and wages could be shared.

In the 1990s the collapse of the Soviet Union and the rapid global spread of capitalism led to further economic crises, especially in the former centrally planned economies. Unemployment, previously unknown, became widespread and in most countries was worse for women than men. Natural disasters, such as drought and floods, caused major development setbacks and civil strife in East Timor, Sri Lanka, the Balkans and several African countries and undermined much of the progress made during the 1970s (Plate 9.1).

Formal employment grew again in the 1990s, particularly in the service industries, although women's jobs tend to enjoy less social protection and employment rights than do men's jobs. Women also increased their share as self-employed workers and as workers in managerial and administrative positions in the 1990s (Elson 2000). Women have been particularly affected by the industrial restructuring brought about as a result of the introduction of new less labour-intensive technology, and of decreased foreign direct investment in assembly industries in most of the 40 developing countries operating export processing zones. Many women joined the informal economy or were forced to migrate. In the Philippines the number of women seeking jobs overseas increased by 70 per cent between 1982 and 1987, and continued to expand in the 1990s (Momsen 1999). Whether women work often depends on the availability of a daughter to take on domestic chores and childcare (Plate 9.2). Girls may be taken out of school to replace the mother and so lose their chance to be trained for a better job in the future, while families will struggle to educate boys (Box 9.1) (Narayan *et al.* 2000b).

Increased food prices force poor families to reduce both the quantity and quality of their food intake and women and children are usually

Box 9.1

The gendered impact of structural adjustment on education levels in northern Ghana

In Zorse, a village of about 2,000 people in north-eastern Ghana, few women are educated, but the number doubled between 1984 and 1991, although the level of education fell. Two-thirds of those with primary education in 1984 had some post-primary schooling, but by 1991 the proportion had fallen to 17 per cent, indicating an increasing school drop-out rate for girls. For Ghanaian girls the national primary school enrolment rate fell 3.4 per cent between 1982 and 1989 as a result of the introduction of school fees at all levels as part of the structural adjustment programme.

In Zorse, primary school fees were introduced in 1986. In addition, a chair, desk, books and stationery, which used to be provided by the government, had now to be provided by parents. School uniform was also a parental responsibility. The total yearly costs for attending primary school were estimated to be £10 in 1991, when the average annual income of women was about £12. Although some men do contribute to the cost of educating their children, traditionally this is the sole responsibility of mothers. Older siblings are expected to help with school fees, but at a time when the cost of living was rising and most people were facing declining real incomes such help had become more limited. The imposition of school fees and other costs is forcing parents to make a choice as to which children to send to school, and this choice is often in favour of boys because of the cultural values in this predominantly Muslim region.

There were very few community organizations in Zorse in 1991, so women's survival strategies were individualized and depended on increasing income-earning work and food production, while utilizing female household members and kinship linkages to cope with the greater workloads. Thus increasing dependence on help from daughters was another reason to take them out of school. Grandmothers also increased income-earning activities in addition to their traditional role of childcare. Over half (52.2 per cent) of the families in the village recorded a decline in food consumption, particularly in terms of protein-rich foods, such as meat and pigeon peas, between 1984 and 1991, while only 2 per cent felt family food intake had improved.

A levy on users of the borehole in the centre of the village, to pay for its maintenance, was introduced in 1986 and from 1991 fees were charged for each container of water taken from the borehole. The borehole had been dug in 1985 and had reduced both the amount of time women spent in water collection and the incidence of water-borne diseases. The introduction of charges forced poor women to turn once again to the more distant polluted streams and unlined well they had used previously. Such changes increase the gap between the poorest and the less-poor families in the village in terms of workload, health status and educational levels.

With the majority of women being poor rural producers, the cost recovery element of the structural adjustment programme threatened the success of the national educational reforms undertaken with the aim of increasing access to schools. It was also expected to lead to further gender differences in educational achievements which eventually translate into differences in employment opportunities.

Source: Awumbila (1994: 254–60).

most affected. Malnutrition combined with lack of adequate health care is being reported from some of the former communist countries (ibid.). When families can no longer cope, children are abandoned to fend for themselves on the streets. The street children seen in Latin America in the 1980s had appeared in Africa and some transition countries by the late 1990s (ibid.). Men migrate to seek work elsewhere, with the result that the number of female-headed households increases. However, in the last decade young women have begun to migrate independently in search of employment often in domestic service or sex work.

Development policies are now aimed at sustainable pro-poor growth. Progress towards this is being assessed through success at achieving the targets of the Millennium Development Goals (MDGs) (UNDP 2003). The first of the millennium goals is to reduce by half the proportion of people living in extreme poverty by 2015 (World Bank 2001). Of the world's six billion people, 1.2 billion live on less than US$1 per day (ibid.). Global poverty rates declined by 20 per cent during the 1990s and the number of people living in extreme poverty almost halved, but most of this decline occurred in China and India, while most other parts of the world are falling short of the poverty target. At the same time aid to developing countries fell: in Africa it declined in real terms over the decade from US$39 to US$19 per capita. In order to reach the first MDG, high rates of economic growth are necessary but unlikely to occur, and the number of very poor people living in sub-Saharan Africa is expected to rise. This situation is exacerbated by the unequal distribution of income in many African countries, of which the worst is Swaziland, where in 1994 half the national income was consumed by the top 10 per cent of the population, while the bottom 30 per cent received less than 4 per cent (World Bank 2001). According to the most recent figures (UNDP 2003), over half the population of Burkina Faso, the Central African Republic, Mali, Niger, Nigeria, Sierra Leone and Zambia was living on less than US$1 a day. Indicators of income-poverty are not calculated in a gender-sensitive way and so there is no way of estimating the relative proportions of poor women and men.

Poverty is not just shortage of material assets but is marked by multiple deprivations. Ill-being and feelings of powerlessness and insecurity are accompanied by troubled and unequal gender relations (Narayan *et al.* 2000b). Increased economic hardship and growing male unemployment is pushing poor women to work outside the home in greater numbers, but this does not automatically give them greater status or security in the home. Women feel overburdened by work and men feel humiliated by being unable to maintain their status as the main breadwinner. These shifts in gender identities are a source of deep anxiety for both women and men.

The other Millennium Goals of reducing gender inequalities in education, infant and child death rates, decreasing maternal mortality rates, improving access to reproductive health services and reducing the incidence of major diseases by 2015 are least likely to be met in the rural areas of sub-Saharan African countries. Sahn and Stifel (2003), using time series data from household surveys for 24 African countries, found a discouraging picture: in only two countries were rural areas on target for poverty reduction, though the urban sectors of five countries were on target. Only the urban areas of three countries and the rural of two, of the ten for which data was available, have witnessed reductions in gender discrimination in enrolment in primary school. No country of the 24 was on target for reducing infant and maternal mortality rates in rural areas and only four were on target in urban areas. In terms of access to family planning a much higher proportion of women knew about modern contraceptive methods than had access to them, with the biggest gap being in rural Burkina Faso, where, although 61 per cent of women knew about modern contraceptive methods, only 2.4 per cent had access to them in 1992 (ibid.). By 1999 90 per cent of women in urban areas and over 70 per cent of women in rural areas (except in Nigeria, Mali and Madagascar) of all 16 countries with data, knew about modern contraceptive methods, but only in Zimbabwe (54 per cent), Kenya (37 per cent) and Tanzania (22 per cent) was the national rate of use of such methods above 20 per cent (ibid.). However, although eight countries are not on target to meet any of the Millennium Goals and ten are not likely to meet them in their rural areas, with only Ghana and Tanzania likely to meet more than one MDG nationally, most show some statistically significant improvement in levels of well-being. Only six (Benin, Burundi, Comoros, Malawi, Mozambique and Rwanda) show no such improvement between the late 1980s and 1990s. These six failing countries have suffered from civil disturbances which have undermined progress. Of the remaining three-quarters of the 24 countries studied, three (Ghana, Madagascar and Niger) have made progress at the national level on five indicators of well-being (ibid.).

Another study of an African country, Ethiopia, takes distributional impacts of a pro-poor development policy into account (Bigsten *et al.* 2003). It was found that completion of primary education by both spouses was important in terms of enabling higher household expenditures, especially in towns, and urban households, in which either husband or wife had at least primary education, had a lower probability of falling into poverty, based on data collected in 1994 and 1997. Lack of education and high dependency ratios were less likely to lead to poverty in rural areas, probably because education is less of a necessity for agriculture than for urban jobs and children and the elderly can contribute more labour in rural areas than in cities (ibid.). These urban location benefits were also found in rural areas close to major towns. A similar pattern supporting the benefits of education for poverty alleviation in urban areas, especially for women, was noted in a study of Malawi (Mukherjee and Benson 2003). Female-headed households, in general, had a higher chance of being among the 'chronically' poor and a lower chance of being among the permanent non-poor, but if the woman household head had primary education she was less likely to be among the 'chronic' poor.

Women become more powerful in the face of such acute threats to the survival of the family because of their traditional responsibility for reproduction. Both mothers and daughters work longer hours and time becomes their scarcest resource (Table 9.1). Women increase their productive work by seeking alternative sources of income to compensate for declines in household income, while also spending longer searching for and preparing cheaper types of food (Moser 1992). Men feel themselves marginalized and often adult males respond by increasing their alcohol consumption and their level of violence to women, while teenage sons turn to dependence on drugs (Narayan *et al.* 2000b). The poorest families, often headed by women, usually bear a disproportionate share of the burden of poverty and economic crises tend to exacerbate pre-existing gender inequalities. Craddock (2001) argues that declining nutrition levels, the increased cost of health services as a result of adjustment policies, and an increased dependence on commercial sex as a source of income made East African countries more vulnerable to the HIV/AIDS epidemic. Overall, the feminization of labour and the growth of the informal economy reflect a weakened position for men rather than greater economic opportunities for women (Narayan *et al.* 2000b). Income earning by women does not necessarily lead to social empowerment or greater gender equity.

Community management

Awareness of the needs of their communities tends to be greater among women than men, since it is normally women who have to cope with problems of housing and access to services. Consequently, women often take the lead in demanding improvements in urban services. They may also work together to change social attitudes. Groups of women in Bombay, India, march silently, carrying placards, around houses in which dowry deaths have occurred, bringing public shame on the perpetrators. This is more effective than government legislation in reducing the number of these tragedies. Women's groups in India have also lobbied for legal restraints on the abortion of female foetuses. As pressure on women's time increases, their community management role may change.

Women's survival strategies often depend on building up networks of women within the community. Such networks may include the extended family as well as colleagues from work and may reach outside the immediate community. Communal kitchens in Lima, Peru, have spread rapidly and by the mid 1990s 2000 *comedores populares* were feeding about 200,000 daily, ensuring minimal nutritional subsistence for people in local communities (Hays-Mitchell 2002). Such organizations also free up time for individual women by relieving them of some domestic chores and provide a locus for broader community action. Development agencies often advocate the spread of these grass-roots separate women's organizations because they feel that they avoid confrontation with cultural patterns, which oppose the mixing of unrelated women and men, and prevent a submergence of women's interests and loss of leadership to men. These women's groups may provide a focus for the politicization of women's lives around issues of prime importance to their domestic role, such as rising food costs and the disappearance of their children at the hands of repressive regimes. In this way, organizing to meet practical gender needs may lead to political activism to achieve strategic gender needs (Waylen 1996).

This link between the empowerment of women for household welfare and consequent political action has not been analysed by most development workers. An important example of such empowerment is the Red Thread movement in Guyana. This began in 1986 as a group of craft workers, with leadership from middle-class women political activists, aimed at improving employment opportunities and access to resources for poor women. It developed into a political force, uniting women across class and racial boundaries who were involved in consciousness raising and community political action (Nettles 1998).

As governments are forced to cut back on public sector spending, the burden of providing basic needs services to poor communities is falling increasingly on women, for example women in Argentina have organized soup kitchens (Narayan and Petesch 2002). Women are having to spend precious time negotiating directly with international agencies and non-governmental organizations in order to get free assistance for their children, and training or credit for setting up income-earning enterprises (Moser 1992). Narayan and Petesch (2002) note that poor people often find NGOs more helpful than state agencies and that the education provided by NGOs to poor women in Bangladesh has made them more aware of their legal rights.

Development projects directed at women are often small, scattered and peripheral to the main aims of development. They usually try to promote greater self-sufficiency rather than development in the sense of expansion and qualitative change. Furthermore, the criteria for success are often less stringent than those for projects specifically for men. Projects aimed at women, such as microfinance, are often based on groups but these artificial groups may undermine other groups already in existence and so reduce the role of women in the local economy, as occurred among fishing communities in coastal Ghana (Walker 1998). On the other hand, when general development projects are planned women may find themselves excluded because of restrictive entry conditions. Female-headed households are numerous in the urban slums of the South, but eligibility for new housing is commonly based on the premise that there is a male head of household (Moser and Peake 1987). Female household heads may not have an income which is large enough or secure enough to qualify for housing. In self-help projects they may not have the time or skills necessary to build a house and if they employ a man to do it they may be cheated. These problems are now widely recognized and can be overcome through training, membership of a women's group or special eligibility conditions tailored to suit the constraints of women's lives.

The success of the women who organized the rebuilding of their homes after the earthquake in Mexico City in 1986 is a good example. They learned how to lobby politicians, how to design apartments which were suitable for their lifestyles and family size and how to prevent contractors cheating them. In doing so they not only rehoused their families but also successfully challenged the patriarchal structure of households, trade unions and political parties. In the face of an environmental disaster, grass-roots women's groups were instrumental in reviving their communities.

Gender and the new economy

The new economy is characterized by globalization and the increasing use of computing and information technologies, but also by deregulation, income polarization and feminization of employment, with new more flexible patterns and hours of work. Globalization, whereby producers and investors behave as if the world is a single market linked by flows of capital and goods rather than a set of individual national economies, has been thought to offer both opportunity and difficulties for women and men in poor countries. It has been associated with environmental degradation and human exploitation, as well as with access to new technology, information and new types of work (Sweetman 2000). Freeman (2001) argues that we have created a new dichotomous model in which globalization is gendered, with the global seen as masculine and the local mapped as feminine. The markets in the present economy are creating new opportunities but distributing them unevenly. Illiterate poor women in isolated rural areas remain untouched by these changes. The ability to grasp the new opportunities is determined by women's and men's different degrees of freedom to undergo training and take up paid work. As women in the South are disproportionately likely to be least educated and, even where educated as in the transition countries, are restricted by their household responsibilities for unpaid reproductive work, they are least likely to benefit from these opportunities (Pearson 2000). This growing gender gap of opportunity results in a new feminization of poverty.

For some younger women, globalization has created new jobs in export-oriented industries, such as garment-making and electronics manufacturing, especially in such areas as the United States/Mexican border region and in Sri Lanka, Bangladesh and Malaysia. For the more educated women there are new opportunities in call centres and data entry, especially where English is spoken, as in many parts of the Caribbean, where almost 5,000 women were employed in data entry in the late 1990s (ILO 2001; Freeman 2001). Women with higher education have been able to benefit from the global spread of financial institutions, especially in Eastern Europe, where women are more likely than men to be trained in economics and to speak several languages. They are also in demand as computer programmers. These jobs have empowered young women by providing an income and exposing them to new cultural concepts and ways of living (Box 9.2) (Buang 1993; Wolf 1992). Work in data processing, although it may be better than other jobs available locally, may not lead to upgrading in the labour market, and may be seen as a new kind of sweatshop (Pearson 1993).

Access to mobile telephones has allowed women to set up Village Phone Centres in Bangladesh, initiated by the Grameen Bank, and in India, where over 250,000 jobs have been created since 1997 (ILO 2001). Such entrepreneurial activity provides income for the mainly women operators and enables villagers to keep in touch with migrant relatives and friends, but has only weak linkages to the rural poor

Box 9.2

ICT in India: gendered impacts in the formal and informal sectors

In 2002 India began to regain outsourcing contracts with the USA and Britain for back-office services. India has a large, well-educated English-speaking workforce and wages are much lower than in the USA. Delta Airlines moved its wordwide reservations services to Bangalore and expected to save US$12–15 million annually. Such jobs attract young urban men and women college graduates. They work mainly at night in order to synchronize their work schedules with those in the USA. One young woman, Saranya Sukumaran, after three months on the job and at the age of 21, was taking home US$200 a month for her work on the help desk of an American Internet service provider. She works from 10.30 p.m. to 7.30 a.m. five days a week, and calling herself 'Sharon', she responds to calls routed to Bangalore. Her wages help her father, who earns only one-third more than his daughter, to support their household. But Saranya is also able to buy little luxuries for herself and says that she is now her own Santa Claus.

At the other end of the income spectrum, in a country where almost half the population lives on less than one dollar a day, the new technology is also bringing benefits. SEWA, the Self-Employed Women's Association, a credit union for women, has been organizing women in the informal sector in India since 1972, has a membership of 215,000 and has promoted 84 cooperatives and federations of women to access mainstream markets. SEWA has been a leader in India in seeing the potential of ICT for the informal sector. It has carried out capacity building measures by running computer-awareness programmes and by teaching basic computer skills to its team leaders, 'barefoot' managers and members of its various associations. Many of SEWA's member organizations have developed websites and sell their products in the global virtual market place. Computerization of stocks and inventory has speeded up management and has cut customer turnaround time by half.

Making computers and training available is a necessary but not sufficient element for the spread of ICT. The entire membership of SEWA consists of poor self-employed women, so getting access to software in local languages is very important if the benefits of ICT are to be widely available at the grass-roots. Furthermore, two-thirds of SEWA members live in rural areas, where reliable electricity supplies and air-conditioning are rare. Gradually, however, village women are becoming aware of the outside world and are being offered simple computer-based training programmes in such topics as nutrition and healthcare, in order to help improve their own and their families' lives.

Source: adapted from Rai (2002) and ILO (2001: 195).

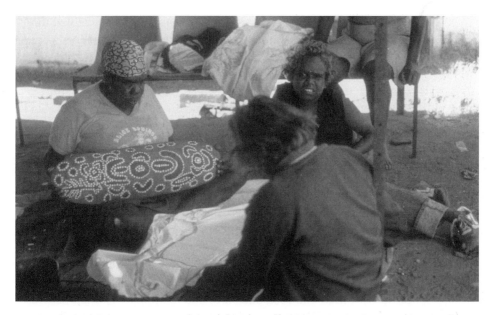

Plate 9.3 *Australia: Aborigine painter. The Australian government at first ignored women's land rights in aboriginal areas. The women formed protest groups and forced planners to reconsider gender differences in the traditional importance of specific sites. Market demand for aboriginal paintings has created new income-earning opportunities for women in this traditionally male-dominated activity.*
Source: Elspeth Young (1942–2002), formerly of the Australian National University, Canberra, ACT

who largely remain outside the information loop (ibid.). The Swaminathan Foundation's Village Knowledge Centres provide ICT facilities within project villages in southern India, including dedicated websites with a variety of locally relevant content. These have significantly increased local knowledge of good nutrition and medical practice, but do not necessarily translate into a meal or substitute for contaminated water supplies and have limited effect on poverty alleviation (ibid.). Computers are also now being used in many poor countries to find global markets for artisanal items and products from small workshops (Momsen 2001) (Box 9.2) (Plate 9.3). Gajjala (2002) discusses the need to create women-friendly technological environments which allow both men and women to be empowered and allow women to combine ICT and domestic duties. Efforts by women worldwide to place gender firmly at the centre of the World Summit on the Information Society stress the importance of enhancing the possibilities for women at all levels to use ICT tools for the transformation of gender hierarchies in society, and to challenge stereotyped gender roles (Walker and Banks 2003).

The spread of global communications technologies has enabled the highly influential international women's movement to pioneer a global advocacy campaign, including but also going beyond the series of United Nations conferences in the 1990s. In this way, marginalized groups have unprecedented opportunities to create pressure groups that may ultimately transform national politics. Such networking has enabled the voices of isolated people to be heard and has facilitated international pressure on governments to stop cruel punishments, such as being stoned to death, as was threatened for a rape victim in northern Nigeria, or to improve accessibility to medical treatment for mothers with AIDS in South Africa.

Conclusion

Economic crisis in many poor countries, enhanced by their peripheral position in the world economy, has led to reductions in spending on health, education and food subsidies and the impact is often heaviest on poor women. When women are able to respond successfully to crises they gain status within the household, either because they have become the chief income earner in the family, or because they have gained confidence through learning how to negotiate successfully with national and international agencies, and how to work with other women. This very success may provoke an additional crisis in the internal gender relations of the household. Women's increased power and independence may result in a male backlash of violence and the expansion of female-headed households. It may also lead to more equality and freedom of choice for both men and women. The conflict between patriarchy, or male dominance, and economic need is creating societies in a state of flux in many parts of the world.

Ways must be found to reduce the gender inequalities in the burden of work if both women and men are to realize their potential. Development plans for women, where they exist, tend to assume mistakenly that women have free time to devote to new projects and also tend to ignore the heterogeneity and differentiation of women. New approaches to development focus on community and participatory development but great care needs to be taken to make sure that the voices of all are heard, especially of women and ethnic minorities.

People of the South and the transition countries are agents of change, not just victims. The United Nations has realized that the role and status of women are central to changes in population and development. It now argues that development plans must be rethought from the start so that gendered abilities, rights and needs are taken into account at every stage. Making investment in women a development priority will require a major change in attitudes to development, not only by governments but also by lending agencies.

After three decades of Women in Development and Gender and Development policies the work of redressing gender inequalities has only just begun. It is often said that women constitute most of the poorest of the poor, but there is no empirical evidence for this statement (Momsen 2002a; Elson 2000) and blaming household poverty on the growth of female-headed households is also unproven (Momsen 2002a). The gender gap in wages in industry and services has fallen between 1980 and 1997 in more countries and increased in fewer but globalization makes it more difficult for individual countries to reduce the gender differential (Elson 2000). Investing in women is not a global panacea. It will not put an end to poverty but it will make a critical contribution to improving household well-being. Furthermore, it will help to create the basis for future generations to make better use of both resources and opportunities.

Development aims at both economic betterment and greater equity but how these two concerns should be achieved simultaneously has yet to be agreed. The concept of human rights is increasingly being seen as the key. Consideration of human rights in development usually incorporates rights to an equal voice, information, political participation and public accountability, as well as the equal right to access to material benefits, such as clean water, land, education, food, housing, health, credit and employment. The Millennium Goals are attempting to deliver some of these rights but are already recording spatial differences in achievement. As the examples in this book show, gender balance in human rights is hard to deliver. States may pass laws providing equal access to women and men to property rights but these laws may not be enforced at the grass-roots level, or may be superseded by customary laws. Efforts to improve gender equity may lead to backlash from those who see themselves losing their superior position. Above all, attempts to introduce gender equity in public spaces may be undermined by strong patriarchal behaviour in the domestic sphere as illustrated in Box 4.3. Levels of gender equity will also vary by class, race, religion, citizenship and individual beliefs in any particular state. The Vienna Conference on Human Rights in 1993 urged that Governments and the United Nations should ensure 'the full and equal enjoyment by women of all human rights', and emphasized the importance of the full participation of women 'as both agents and beneficiaries' of development (United Nations 1996: 60). These ideals were reinforced at the Cairo Conference on

Population and Development in 1994, the Copenhagen World Summit for Social Development in 1995 and the Fourth World Conference on Women in Beijing in 1995. Women's rights as human rights and gender equity in development are on the international agenda. We must continue to actively monitor progress if women and men are to participate in and benefit equally from development.

Learning outcomes

- There has been progress in gender equality as measured by the Gender Development Index.
- Many countries have introduced quotas to improve the gender balance in politics and the effect can be seen in policy-making, especially at the local level.
- Poverty forces people to increase their involvement in the market economy, to migrate, to cut back on education for their children and to seek a range of survival strategies.
- The new economy has provided new types of employment and networking opportunities for some people, but the uneducated, especially older women, are often marginalized.

Discussion questions

1   Explain how women's networks and women local political leaders can play an important role in grass-roots projects for community improvement and social change.

2   Distinguish between the practical and strategic needs of women.

3   Why are overwork and shortage of time often ignored as barriers to women's participation in development projects?

4   To what extent has information and communications technology (ICT) provided access to knowledge and global linkages for poor people in isolated parts of the developing world?

Further reading

Afshar, Haleh and Carolyne Dennis (eds) (1992) *Women and Adjustment Policies in the Third World*, Women's Studies at York, Macmillan Series, Basingstoke: Macmillan. A collection of papers on the gendered impact of structural adjustment by some of the leading scholars and activists in the field.

Beneria, Lourdes and Shelley Feldman (eds) (1992) *Unequal Burden: Economic Crises, Persistent Poverty, and Women's Work*, Boulder, CO and Oxford: Westview Press. Based on a workshop held in 1988, this book looks at the impact of structural adjustment on women's work. The book has two useful overview chapters by Shelley Feldman and Diane Elson, plus three case studies on Latin America, two on South Asia and one each on the Caribbean and on urban Tanzania.

Fenster, Tovi (ed.) (1999) *Gender, Planning and Human Rights*, London and New York: Routledge. This book includes chapters on both North and South and post-communist countries. The contributors consider human rights in relation to migration, indigenous and minority peoples, domestic violence and microcredit from a feminist perspective.

Jackson, Cecile and Ruth Pearson (1998) *Feminist Visions of Development: Gender Analysis and Policy*, London and New York: Routledge. A post-Beijing analysis of the current state of gender and development issues in relation to policy.

Marchand, Marianne H. and Anne Sisson Runyan (eds) (2000) *Gender and Global Restructuring: Sightings, Sites and Resistances*, London and New York: Routledge. Twelve case studies from both countries of the South and those in transition.

Pearson, Ruth (2000) '"Moving the goalposts": gender and globalization in the twenty-first century', *Gender in the 21st Century*, Oxford: Oxfam GB. Useful overview article. This volume has several relevant articles and an editorial by Caroline Sweetman.

Radcliffe, Sarah A. and Sallie Westwood (1993) *'Viva': Women and Popular Protest in Latin America*, London and New York: Routledge. A collection of case studies of gendered political activism in Latin America.

Waylen, Georgina (1996) *Gender in Third World Politics*, Buckingham: Open University Press. Presents an analysis of the links between gender, political systems and development.

Websites

www.ids.ac.uk/bridge/index.html   BRIDGE (Briefings on development and gender). This is an innovative information and analysis service provided by the Institute of Development Studies at the University of Sussex. It specializes in gender and development. The papers can be downloaded.

www.un-instra.org/en/research/gender_and_ict/docs   United Nations International Research and Training Institute for the Advancement of Women (INSTRAW) is based in the Dominican Republic. It carries out data gathering and research on gender issues. In 2002 it hosted a Virtual Seminar Series on Gender and ICTs.

www.wigsat.org   Women in Global Science and Technology (WIGSAT) aims to facilitate global networking among women scientists and technologists on critical issues in science and technology.

References

Abasiya, Gifty (2002) 'Call for stronger women's role in conflict resolution'.
  Available on-line: http://www.allAfrica.com (accessed 25 November
  2002).
Abildtrup, Ulla and Peter Rathmann (2002) 'Country report: Lesotho', *The
  Courier ACP-EU* 194: 70–92.
Afshar, Haleh (ed.) (1991) *Women, Development and Survival in the Third
  World*, London and New York: Longman.
Afshar, Haleh and Stephanie Barrientos (eds) (1999) *Women, Globalization and
  Fragmentation in the Developing World*, London and Basingstoke:
  Macmillan; New York: St Martin's Press.
Afshar, Haleh and Carolyne Dennis (eds) (1992) *Women and Adjustment
  Policies in the Third World*, Women's Studies at York, Macmillan Series,
  Basingstoke: Macmillan.
Agarwal, B. (1992) 'The gender and environment debate: lessons from India',
  Feminist Studies 18 (1): 119–58.
Agarwal, B. (1997a) 'Gender, environment and poverty interlinks: regional
  variations and temporal shifts in rural India, 1971–91', *World Development*
  25 (1): 23–52.
Agarwal, B. (1997b) 'Re-sounding the alert: gender, resources and community
  action', *World Development* 25 (9): 1373–80.
Aggleton, Peter and Richard Parker (2002) 'World AIDS Campaign 2002–2003:
  a conceptual framework and basis for action: HIV/AIDS stigma and
  discrimination', Geneva: UNAIDS. Available on-line:
  http://www.unaids.org/publications/documents/human/JC781-Concept
  Framew-E.pdf (accessed 15 October 2002).
Ahmed, M. Feroze (ed.) (2000) *Bangladesh Environment 2000*, Proceedings of
  the International Conference on Bangladesh Environment 2000, Dhaka,
  Bangladesh, 14–15 January, 2000, pp. 84–229.

Akhter, Farida (2000) 'Micro-credit: the development devastation for the poor'. Available on-line: http://www.uea.ac.uk/dev/greatnet (accessed 3 November 2000).

Aladuwaka, Seela (2002) Unpublished field notes.

Anderson, Bridget (2000) *Doing the Dirty Work: The Global Politics of Domestic Labour*, London: Zed Books.

Annan, Kofi (2002) '"The Impact of Violent Conflict on Women and Girls": study released today by UN Secretary General', New York: Women's Global Net, 21 October. Available on-line: http://www.#211:iwtc@iwtc.org (accessed 1 November 2002).

Appadurai, Arjun (1996) *Modernity at Large: Cultural Dimensions of Globalization*, Minneapolis: University of Minnesota Press.

Ardayfio-Schandorf, Elizabeth (1993) 'Household energy supply and women's work in Ghana', in Janet H. Momsen and Vivian Kinnaird (eds) *Different Places, Different Voices: Gender and Development in Africa, Asia and Latin America*, London: Routledge.

Ardayfio-Schandorf, Elizabeth (forthcoming) 'Rural energy supply and women's work in Ghana', in Janet H. Momsen (ed.) *Geographical Studies in Women and Development* (provisional title), London: Routledge.

Aubel, Judi, Ibrahima Touré, Mamadou Diagne, Kalala Lazin, El Hadj Alioune Sème, Yirime Faye and Mouhamadou Tandia (2001) 'Strengthening grandmother networks to improve community nutrition: experience from Senegal', *Gender and Development* 9 (2): 62–73.

AWID Resource Net (2001) *Friday File, Issue 51*. On-line posting. Available e-mail: awid@awid.org (19 November 2001).

AWID Resource Net (2002a) *Friday File, Issue 73*. On-line posting. Available e-mail: awid@awid.org (19 April 2002).

AWID Resource Net (2002b) *Friday File, Issue 117*. On-line posting. Available e-mail: awid@awid.org (26 June 2002).

AWID Resource Net (2002c) *Announcements, Issue 13*. On-line posting. Available e-mail: contribute@awid.org (27 November 2002).

Awumbila, Mariama (1994) 'Women and change in Ghana: the impact of environmental change and economic crisis on rural women's time use', unpublished thesis, University of Newcastle upon Tyne.

Awumbila, Mariama (2001) 'Infant nutrition practices in northern Ghana', lecture at University of California, Davis, May.

Awumbila, Mariama and Janet H. Momsen (1995) 'Gender and the environment: women's time use as a measure of environmental change', *Global Environmental Change* 5 (4): 337–46.

Bardhan, K. and S. Klasen (1999) 'UNDP's gender-related indices: a critical review', *World Development* 27: 985–1010.

Bardhan, K. and S. Klasen (2000) 'On UNDP's revisions to the gender-related development index'. Available on-line: http://www.hdr.undp.org/docs/statistics/undp-revisions-gender-related.pdf (accessed 20 December 2002).

Barrett, H. R. and A. W. Browne (1989) 'Time for development? The case of women's horticultural schemes in rural Gambia', *Scottish Geographical Magazine* 105 (1): 5.

Bearak, Barry (2002) 'Sip of death: arsenic ravages Bangladesh', *Sacramento Bee*, 14 July: A11.

Begum, Rasheda (1993) 'Women in environmental disasters: the 1991 cyclone in Bangladesh', *Focus on Gender* 1 (1): 34–9.

Bell, Emma (ed.) (2002) 'Gender and HIV/AIDS: *INBRIEF* Issue 11', Sussex: Institute of Development Studies. Available on-line: http://www.ids.ac.uk/bridge/dgb11.html (accessed 27 September 2002).

Beneria, Lourdes and Shelley Feldman (eds) (1992) *Unequal Burden: Economic Crises, Persistent Poverty, and Women's Work*, Boulder, CO and Oxford: Westview Press.

Beneria, Lourdes and Martha Roldan (1987) *The Crossroads of Class and Gender: Industrial Homework, Subcontracting and Household Dynamics in Mexico City*, Chicago: University of Chicago Press.

Benton, Jane (1993) 'The role of women's organisations and groups in community development: a case study from Bolivia', in Janet H. Momsen and Vivian Kinnaird (eds) *Different Places, Different Voices: Gender and Development in Africa, Asia and Latin America*, London and New York: Routledge, pp. 230–42.

Bhuiyan, Rejuan Hossain (2000) 'Social and pyschological scenario of arsenicosis patients: a gender perspective', in Nasreen Ahmad and Hafiza Khatun (eds) *Disaster Issues and Gender Perspectives: Conference Proceedings*, Dhaka: Bangladesh Geographical Society, University of Dhaka, pp. 101–21.

Bigsten, Arne, Bereket Kebede, Abebe Shimeles and Mekonnen Taddesse (2003) 'Growth and poverty reduction in Ethiopia: evidence from household panel surveys', *World Development* 31 (1): 87–106.

Blocker, T. Jean and Douglas Lee Eckberg (1989) 'Environmental issues as women's issues: general concerns and local hazards', *Social Science Quarterly* 70 (3), 586–93.

Bode, Birgitta (1993) 'Structural and actor-oriented analysis of village political process in the Almora District of Kumaon, Uttar Pradesh, India', unpublished MS thesis, University of California, Davis.

Bortei-Doku Aryeetey, Ellen (2002) 'Behind the norms: women's access to land in Ghana', in C. Toulmin, P. Lavigne Delville and Samba Traoré (eds) *The Dynamics of Resource Tenure in West Africa*, Ghana: International Institute for Environment and Development. Available in short form on-line: http://www.id21.org/society/S6aea1g1.html (accessed 18 February 2003).

Boserup, Ester (1970) *Women's Role in Economic Development*, London: Allen and Unwin.

Boutros Boutros-Ghali (1996) 'Introduction', in United Nations (eds) *The United Nations and the Advancement of Women 1945–1996*, New York: United Nations.

Boyle, Paul (2002) 'Population geography: transnational women on the move', *Progress in Human Geography* 26 (4): 531–43.

Bradshaw, Sarah (2001) 'Reconstructing roles and relations: women's participation in reconstruction in post-Mitch Nicaragua', *Gender and Development* 9 (3): 79–87.

Braidotti, R., Ewa Charkiewicz, Sabine Häusler and Saskia Wieringa (1994) *Women, the Environment and Sustainable Development: Towards a Theoretical Synthesis*, London: Zed Books.

Bru-Bistuer, Josepa (1996) 'Spanish women against industrial waste: a gender perspective on environmental grassroots movements', in Dianne Rocheleau, Barbara Thomas-Slayter and Ester Wangari (eds) *Feminist Political Ecology: Global Issues and Local Experiences*, London and New York: Routledge.

Buang, Amriah (1993) 'Development and factory women: negative perceptions from a Malaysian source area', in Janet H. Momsen and Vivian Kinnaird (eds) *Different Places, Different Voices: Gender and Development in Africa, Asia and Latin America*, London: Routledge.

Bulatao, Rudolpho A. (2000) 'Introduction', in Rudolpho A. Bulatao and John B. Casterline (eds) *Global Fertility Transition*, New York: Population Council. (A supplement to *Population and Development Review* 27: 1–14.)

Calio, Sonia Alves (1990) 'The Brazilian economic crisis and its impact on the lives of women', *Political Geography Quarterly* 9 (4): 415–23.

Campbell, Connie, with the Women's Group of Xapuri (1996) 'Out on the front lines but still struggling for voice: women in the rubber tappers' defense of the forest in Xapuri, Acre, Brazil', in Dianne Rocheleau, Barbara Thomas-Slayter and Ester Wangari (eds) *Feminist Political Ecology: Global Issues and Local Experiences*, London and New York: Routledge.

Carroll, Rory (2002a) 'Africa's ugly sisters leave trail of death', *Guardian Weekly*, 7–13 November: 4.

Carroll, Rory (2002b) 'Dirty water, broken pumps and a legacy of disease', *Guardian Weekly*, 5–11 December: 25.

Casinader, Rex, Sepalika Fernando and Karuna Gamage (1987) 'Women's issues and men's roles: Sri Lankan village experience', in Janet H. Momsen and Janet Townsend (eds) *Geography of Gender in the Third World*, London: Hutchinson; New York: State University of New York Press.

Chacón, Richard (2000) 'Small loans, big hopes: microlending lets Haitians seek credit', *Boston Globe*, 7 October: C1.

Chang, Kimberley A. and L. H. M. Ling (2000) 'Globalization and its intimate other: Filipina domestic workers in Hong Kong,' in Marianne H. Marchand and Anne Sisson Runyan (eds) *Gender and Global Restructuring: Sightings, Sites and Resistances*, London and New York: Routledge, pp. 27–43.

Chant, Sylvia (1984) 'Women and housing: a study of household labour in Querataro, Mexico', in Janet H. Momsen and Janet Townsend (eds) *Women's Role in Changing the Face of the Developing World*, Durham: Women and Geography Study Group of the Institute of British Geographers, Durham University.

Chant, Sylvia (1987) 'Family structure and female labour in Querétaro, Mexico', in Janet H. Momsen and Janet Townsend (eds) *Geography of Gender in the Third World*, Albany: State University of New York Press; London: Hutchinson.

Chant, Sylvia (ed.) (1992) *Gender and Migration in Developing Countries*, London: Belhaven Press.

Chattopadhyay, Raghabendra and Esther Duflo (2002) *Women as Policy Makers: Evidence from an India-wide Randomized Policy Experiment*, Working Paper, Calcutta: Indian Institute of Management.

Chowdhury, Quamrul Islam (ed.) (2001) *Bangladesh State of the Environment 2000*, Dhaka, Bangladesh: Forum of Environmental Journalists of Bangladesh.

Cole, John W. and Judith A. Nydon (1990) 'Class, gender and fertility: contradictions of social life in contemporary Romania', *East European Quarterly* 23 (4): 469–76.

Corlett, Jan L. (1999) 'Landscapes and lifescapes: three generations of Hmong women and their gardens', unpublished Ph.D. dissertation, University of California, Davis.

Cornwall, Andrea and Sarah C. White (2000) 'Introduction: men, masculinities and development: politics, policies and practice', *IDS Bulletin* 31 (2): 1–6.

Craddock, Susan (2001) 'Scales of justice: women, equity and HIV in East Africa', in Isabel Dyck, Nancy Davis Lewis and Sara McLafferty (eds) *Geographies of Women's Health*, London and New York: Routledge.

Croll, Elizabeth (2000) *Endangered Daughters: Discrimination and Development in Asia*, London: Routledge.

Crossette, Barbara (2001) 'UN agency sets its sights on curbing child marriage: data show harm in premature pregnancy', *New York Times*, 8 March: A6.

Daley-Ozkizilcik, Lynette (1993) 'Labor force participation of rural women in the Aegean region of Turkey', unpublished MS thesis, University of California, Davis.

Denny, Charlotte (2000) 'Healing the child victims of Thailand's brothels', *Guardian Weekly*, 6–12 July: 24.

Derbyshire, Helen (2002) *Gender Manual: A Practical Guide for Development Policy Makers and Practitioners*, London: DFID, Social Development Division.

Dicken, Peter (1998) *Global Shift: Transforming the World Economy*, New York and London: Guildford Press.

Drèze, J. and A. Sen (1989) *Hunger and Public Action*, Oxford: Oxford University Press.

Dugger, Celia W. (2000) 'Kerosene: weapon of choice for attacks on wives in India', *New York Times*, 26 December: A1, A10.

Dugger, Celia W. (2001) 'Means to pick baby's sex sets off furor in India', *New York Times*, 23 November: A14.

Dyck, Isabel, Nancy Davis Lewis and Sara McLafferty (eds) (2001) *Geographies of Women's Health*, London and New York: Routledge.

Dyer, Geoff (2002) 'Developing world afflicted by "diseases of affluence"', *Financial Times*, 31 October: 5.

Dyer, Geoff, Edward Luce and James Kynge (2002) 'China and India are on the edge of an Aids epidemic: is it too late to stop Asia reliving Africa's nightmare?', *Financial Times*, 27 November: 12.

ECLAC (Economic Commission for Latin America and the Caribbean) (2002a) 'Latin America and the Caribbean: selected gender-sensitive indicators', *Demographic Bulletin* xxxv (70). Available on-line: http://www.eclac.cl/ publicaciones/Poblacion/2/LCG2172p/boldem70.pdf (accessed 2 December 2002). Available on-line: http://www.eclac.cl/publicaciones (accessed 2 December 2002).

ECLAC (Economic Commission for Latin America and the Caribbean) (2002b) *Social Panorama of Latin America 2001–2002*, briefing paper. Available on-line: http://www.eclac.cl/ publicaciones (accessed 25 November 2002).

Economist (2001) 'Marriage and divorce, Emirates-style', 27 January: 48.

Edwards, Louise (2000) 'Women in the People's Republic of China: new challenges to the grand gender narrative', in Louise Edwards and Mina Roces (eds) *Women in Asia: Tradition, Modernity and Globalization*, Ann Arbor, MI: University of Michigan Press.

Elson, Diane (coordinator) (2000) *Progress of the World's Women 2000*, New York: Unifem.

ESCAP (Economic and Social Commission for Asia and the Pacific) (1999) *Statistics on Women in Asia and the Pacific, 1999*, Bangkok: United Nations.

Evans, Harriet (1992) 'Monogamy and female sexuality in the People's Republic of China', in Shirin Rei, Hilary Pilkington and Annie Phizacklea (eds) *Women in the Face of Change*, London and New York: Routledge, pp. 147–63.

Evans, Ruth (2002) 'Poverty, HIV, and barriers to education: street children's experiences in Tanzania', *Gender and Development* 10 (3): 51–62.

Fairbairn-Dunlop, Peggy (2002) 'Polynesian boys as girls?', personal e-mail, 2 December.

Farley, Maggie (2002) 'Female AIDS cases on rise', *Los Angeles Times*, 27 November: A1.5.

FAO (Food and Agriculture Organization) (1993) *Agricultural Extension and Farm Women in the 1980s*, Rome: FAO.

FAO (Food and Agriculture Organization) (2002) *HIV/AIDS, Food Security and Rural Livelihoods*, World Food Summit Five Years Later Fact Sheet, Rome: FAO.

FAOSTAT (2002) 'Gender and food security statistics'. Available on-line: http://www.fao.org/gender/eu/stats-e.htm (accessed 16 October 2002).

Feldman, Shelley (1992) 'Crisis, Islam and gender in Bangladesh: the social construction of a female labor force', in Lourdes Beneria and Shelley Feldman (eds) *Unequal Burden: Economic Crises, Persistent Poverty, and Women's Work*, Boulder, CO and Oxford: Westview Press, pp. 105–30.

Fenster, Tovi (ed.) (1999) *Gender, Planning and Human Rights*, London and New York: Routledge.

Finnane, Antonia (2000) 'Dead daughters, dissident sons and human rights in China', in Anne-Marie Hilsdon, Martha Macintyre, Vera Mackie and Maila Stivens (eds) *Human Rights and Gender Politics: Asia-Pacific Perspectives*, London and New York: Routledge.

Forbes, Phyllis (1999) *Forum Libre*, Port au Prince, Haiti: publisher not known.

Freeman, Carla (2001) 'Is local:global as feminine:masculine? Rethinking the gender of globalization', *Signs* 26 (4): 1007–37.

Gaard, Greta (1994) 'Misunderstanding ecofeminism', *Z Magazine* 3 (1): 20–4.

Gain, Philip (1998) *The Last Forests of Bangladesh*, Dhaka: Society for Environment and Human Development.

Gajjala, Radhika (2002) 'Cyberfeminist technological practices: exploring possibilities for a woman-centered design of technological environments', UN/INSTRAW Virtual Seminar Series on Gender and ICTs, Seminar Two: Women and ICTs: Enabling and Disabling Environments, 15–26 July. Available on-line: http://www.un-instraw.org/en/research/gender_and_ict/docs (accessed 30 July 2002).

Gannon, N. (2001) 'Women in Afghanistan', *New York Times*, 23 November: 27. (Original Associated Press article, page 5).

Garwood, Shae (2002) 'Working to death: gender, labour and violence in Cuidad Juarez, Mexico', *Peace, Conflict and Development: An Interdisciplinary Journal* 2 (December): 1–19. Available on-line: http://www.peacestudiesjournal.org.uk/docs (accessed 8 February 2003).

Glassman, Jim (2001) 'Women workers and the regulation of health and safety on the industrial periphery: the case of northern Thailand', in Isabel Dyck, Nancy Davis Lewis and Sara McLafferty (eds) *Geographies of Women's Health*, London and New York: Routledge.

Goetz, A. M. and S. Gupta (1996) 'Who takes credit? Gender, power, and control over loan use in rural credit programs in Bangladesh', *World Development* 24 (1): 43–63.

Gollin, L. X. (1997) 'Taban Kenyah: a preliminary look at the healing plants and paradigms of the Kenyah Dayak people of Kayan Mentarang', in K. W. Sorenson and B. Morris (eds) *People and Plants of Kayan Mentarang*, Jakarta: WWF Indonesia Program.

González de la Rocha, Mercedes (1994) *The Resources of Poverty: Women and Survival in a Mexican City*, Oxford, UK and Cambridge, MA: Blackwell.

Graham, Elspeth (1995) 'Singapore in the 1990s: can population policies reverse the demographic transition?', *Applied Geography* 15 (3): 219–32.

Griffith, Allison (2001) 'Water conservation in Barbados', unpublished M.Sc. thesis, Department of Geography and Geology, University of the West Indies, Jamaica.

Harris, Colette (2000) *Control and Subversion: Gender, Islam and Socialism in Tajikistan*, Amsterdam: University of Amsterdam.

Harris, Colette and Ines Smyth (2001) 'The reproductive health of refugees: lessons beyond ICPD', *Gender and Development* 9 (2): 10–22.

Harry, Indra S. (1980) 'Women in agriculture in Trinidad', unpublished M.Sc. thesis, University of Calgary, Canada.

Harry, Indra S. (1993) 'Women in agriculture in Trinidad: an overview', in Janet H. Momsen (ed.) *Women and Change in the Caribbean*, London: James Currey; Indianapolis: Indiana University Press; Kingston, Jamaica: Ian Randle.

Hays-Mitchell, Maureen (2002) 'Resisting austerity: a gendered perspective on neo-liberal restructuring in Peru', *Gender and Development* 10 (3): 71–81.

Herzfeld, Beth (2002) 'Slavery and gender: women's double exploitation', *Gender and Development* 10 (1): 50–6.

Hesperian Foundation (2001) *Women's Health Exchange, No. 7*, Berkeley, CA: Hesperian Foundation.

Hochschild, Arlie R. (1983) *The Managed Heart*, Berkeley: University of California Press.

Hossain, Khondeker Mokaddem (2000) 'The contribution of homestead forests to the rural Bangladesh people during disaster: a sociological study', in Nasreen Ahmad and Hafiza Khatun (eds) *Disaster Issues and Gender Perspectives: Conference Proceedings*, Dhaka: Bangladesh Geographical Society, University of Dhaka, Bangladesh, pp. 372–88.

Howard-Borjas, Patricia (2001) *Women in the Plant World: The Significance of Women and Gender Bias for Botany and for Biodiversity Conservation*, inaugural address, Wageningen, Neth.: Wageningen University.

Huda, Shahnaz (1994) '"Untying the knot": Moslem women's right of divorce and other incidental rights in Bangladesh', *The Dhaka University Studies Part F* 5 (1): 133–57.

Huda, Shahnaz (1997) 'Child marriage: social marginalisation of statutory laws', *Bangladesh Journal of Law* 1 (2): 139–81.

Huda, Shahnaz (1998) 'Custody and guardianship of minors in Bangladesh', in Tahmina Ahmad and Md. M. A. Khan (eds) *Gender in Law*, Dhaka: ADTAM Publishing House.

Hunt, Juliet and Nalini Kasynathan (2001) 'Pathways to empowerment? Reflections on microfinance and transformation in gender relations in South Asia', in C. Sweetman (ed.) *Gender, Development and Money*, Focus on Gender Series, Oxford: Oxfam. Available in abstract on-line: http://www.id21.org.socety/s6cjh1g1.html (accessed 3 February 2003).

Huq-Hussain, Shahnaz (1996) *Female Migrants' Adaptation in Dhaka: A Case of the Processes of Urban Socio-Economic Change*, Dhaka, Bangladesh: Urban Studies Programme, Department of Geography, University of Dhaka.

Huq-Hussain, Shahnaz (2002) Personal communication, May.

Hyma, B. and Perez Nyamwange (1993) 'Women's role and participation in farm and community tree-growing activities in Kiambu District, Kenya', in Janet H. Momsen and Vivian Kinnaird (eds) *Different Places, Different Voices: Gender and Development in Africa, Asia and Latin America*, London: Routledge.

IBGE (Instituto Brasileiro de Geografia e Estatística) (1990) *Estatísticas Históricas do Brasil*, Rio de Janeiro: IBGE.

IBGE (Instituto Brasileiro de Geografia e Estatística) (1994) *Anuário Estatístico do Brasil 1994*, Rio de Janeiro: IBGE.

IBGE (Instituto Brasileiro de Geografia e Estatística) (2002) 'National sample survey of households 1995 and 1999', data organized by the Brazilian Institute of Municipal Administration/Economic and Social Development Area/Nucleus of Women's Studies and Public Policy.

Ibrahim, O. (1987) 'The labour force in Libya: problems and prospects', unpublished Ph.D. thesis, University of Durham, England.

ICRW (International Center for Research on Women) (1996) *Women's Health Matters*, Global Fact Sheet, Washington, DC: ICRW.

ILO (International Labour Organization) (1983) *Yearbook of Labour Statistics*, Geneva: ILO.

ILO (International Labour Organization) (1999) 'Women and training in the global economy', in *World Development Report 1998–99*, Geneva: ILO.

ILO (International Labour Organization) (2001) World Employment Report 2001: Life at Work in the Information Economy, Geneva: ILO.

Insights Development Research (2002) 'Mind the gap: bridging the rural–urban divide', *Insights* 41 (May), Brighton, Sussex: ID21 Institute of Development Studies.

INSTRAW (1999) *Ageing in a Gendered World: Women's Issues and Identities*, Santo Domingo: INSTRAW.

IRINnews (UN Office for the Coordination of Humanitarian Affairs) (2003) 'Kenya: hundreds of girls run away to avoid FGM'. Available on-line: http://www.irinnews.org (accessed 11 February 2003).

Ismail, F. Munira (1999a) 'Multi-ethnic analysis of gendered space amongst rural women in Sri Lanka', Ph.D. dissertation, University of California, Davis.

Ismail, F. Munira (1999b) 'Maids in space: gendered domestic labour from Sri Lanka to the Middle East', in Janet H. Momsen (ed.) *Gender Migration and Domestic Service*, London: Routledge, pp. 229–41.

Itzigsohn, José and Silvia Giorguli Saucedo (2002) 'Immigrant incorporation and sociocultural transnationalism', *International Migration Review* 36 (3): 766–98.

Iyun, B. Folasade and E. A. Oke (1993) 'The impact of contraceptive use among urban traders in Nigeria: Ibadan traders and modernisation', in Janet H. Momsen and Vivian Kinnaird (eds) *Different Places, Different Voices: Gender and Development in Africa, Asia and Latin America*, London: Routledge, pp. 63–73.

Jackson, C. (1994) 'Gender analysis and environmentalism', in T. Benton and M. Redclift (eds) *Theory and the Global Environment*, London: Routledge, pp. 113–49.

Jiggins, J. (1986) 'Women and seasonality: coping with crisis and calamity', *IDS Bulletin* 17 (3), 9–18.

Joekes, Susan and Ann Weston (1994) *Women and the New Trade Agenda*, New York: Unifem.

Johnson, Brooke R., Mihai Horga and Laurentia Andronache (1996) 'Women's perspectives on abortion in Romania', *Social Science and Medicine* 42 (4): 521–30.

Johnson, Noelle (2001) 'The nutritional implications of deforestation: how ecological change is affecting the availability of wild plants as beta sources: a study of two Karen communities in Northern Thailand', unpublished MS thesis, University of California, Davis.

Kabeer, N. (1994) *Reversed Realities*, London: Verso.

Kabeer, N. (1998) 'Money can't buy me love'? Re-evaluation of gender, credit and empowerment in rural Bangladesh', Discussion Paper 363, Brighton, Sussex: Institute of Development Studies.

Khosla, Prabha (2002a) 'Water crisis in the Ukraine', *Women and Environments* 56/57 (Fall): 36–7.

Khosla, Prabha (2002b) 'Women at the Sustainability Summit: shifting parameters of success in a globalizing world', *Women and Environments* 56/57 (Fall): 17–20.

King, Ynestra (1983) 'The ecofeminist imperative', in L. Caldecott and S. Leland (eds) *Reclaim Earth: Women Speak Out for Life on Earth*, London: The Women's Press.

Kinnaird, Vivian and Derek Hall (eds) (1994) *Tourism: A Gender Analysis*, New York and Chichester: John Wiley and Sons.

Kinnaird, Vivian and Janet H. Momsen (1993) 'Introduction', in Janet H. Momsen and Vivian Kinnaird (eds) *Different Places, Different Voices: Gender and Development in Africa, Asia and Latin America*, London and New York: Routledge.

Kirk, Gwyn (1998) 'Ecofeminism and Chicano environmental struggles: bridges across gender and race', in Devon G. Peña (ed.) *Chicano Culture, Ecology, Politics: Subversive Kin*, Tucson, AZ: University of Arizona Press.

Klasen, Stephan and Claudia Wink (2002) 'A turning point in gender bias in mortality? An update on the number of missing women', *Population and Development Review* 28 (2): 285–312.

Kleysen, B. and F. Campillo (1996) *Rural Women Food Producers in 18 Countries in Latin America and the Caribbean*, San José, Costa Rica: IICA.

Kligman, Gail (1992) 'The politics of reproduction in Ceauşescu's Romania: a case study in political culture', *East European Politics and Society*, 6(3): 364–418.

Kligman, Gail (1998) *The Politics of Duplicity: Controlling Reproduction in Ceauşescu's Romania*, Berkeley and Los Angeles: University of California Press.

Knodel, John, Apichat Chamratrithirong and Nibhon Debavalya (1987) *Thailand's Reproductive Revolution: Rapid Fertility Decline in a Third-World Setting*, Madison: University of Wisconsin Press.

Krug, Étienne G., Linda L. Dahlberg, James A. Mercy, Anthony B. Zwi and Rafael Lozano (eds) (2002) *World Report on Violence and Health*, Geneva: World Health Organization.

Lacey, Marc (2003) 'African women gather to denounce genital cutting', *New York Times*, 6 February: A3.

Lamont, James (2001) 'Lesotho seen as gateway to US market', *Financial Times*, 23 August: 7.

Lancaster, John (2002b) 'Brides become scarce in India: sex selection during pregnancy makes for fewer females and many lonely males', *Guardian Weekly*, 12–18 December: 29.

Leach, Melissa, Susan Joekes and Cathy Green (1995) 'Editorial: gender relations and environmental change, *IDS Bulletin* 26 (1): 1–8.

Lefèbvre, François (2002a) 'The struggle for water', *Courier ACP-EU* 194: 13–15.

Lefèbvre, François (2002b) 'Heated talk on energy', *Courier ACP-EU* 194: 15–17.

Lemieux, Gilles (1975) 'Human responses and adjustments to the 1963–65 ashfalls of Irazú Volcano, Costa Rica: a geographical study of environmental perception', unpublished Ph.D. dissertation, Department of Geography, University of Calgary, Canada.

Leslie, Helen (2001) 'Healing the psychological wounds of gender-related violence in Latin America: a model for gender-sensitive work in post-conflict contexts', *Gender and Development* 9 (3): 50–9.

Li, Huey-li (1993) 'A cross-cultural critique of ecofeminism', in Greta Gaard (ed.) *Ecofeminism, Women, Animals and Nature*, Philadephia: Temple University Press.

Lynch, Barbara Deutsch (1997) 'Women and irrigation in highland Peru', in Carolyn E. Sachs (ed.) *Women Working in the Environment*, Washington, DC: Taylor & Francis.

McGregor, Liz (2002) 'Botswana wages "fight to the death" against Aids', *Guardian Weekly*, 11–17 July: 3.

McStay, Jan R. and Riley E. Dunlap (1993) 'Male–female differences in concern for environmental quality', *International Journal of Women's Studies* 6 (4): 291–301.

Manikam, P. P. (1995) *Tea Plantations in Crisis: An Overview*, Colombo, Sri Lanka: Social Scientists Association.

Marchand, Marianne H. and Anne Sisson Runyan (eds) (2000) *Gender and Global Restructuring: Sightings, Sites and Resistances*, London and New York: Routledge.

Marcoux, A. (2002) 'Sex differentials in undernutrition: a look at survey evidence, *Population and Development Review* 28 (2): 275–84.

Masika, Rachel (2002) 'Editorial for special issue on climate change', *Gender and Development* 10 (2): 2–9.

Mason, Karen Oppenheim (2000) 'Gender and family systems in the fertility transition', in Rudolpho A. Bulatao and John B. Casterline (eds) *Global Fertility Transition*, New York: Population Council. (A supplement to *Population and Development Review* 27: 160–76.)

Mattingly, Doreen J. (2001) 'The home and the world: domestic service and international networks of caring labor', *Annals of the Association of American Geographers* 91 (2): 370–86.

Mayoux, Linda (2002) *Women's Empowerment or Feminisation of Debt? Towards a New Agenda in African Microfinance*, London: One World Action. Available on-line: http://www.oneworldaction.org (accessed 20 October 2002).

Mehra, Rekha, Thai Thi Ngoc Du, Nguyen Xuan Nghia, Nguyen Ngoc Lam, Truong Thi Kim Chuyen, Bang Anh Tuan, Pham Gia Tran and Nguyen Thi Nhan (1996) 'Women in waste collection and recycling in Ho Chi Minh City', *Population and Environment: A Journal of Interdisciplinary Studies* 18 (2): 187–99.

Merchant, C. (1992) *Radical Ecology: Global Issues and Local Experiences*, London and New York: Routledge.

Merchant, Kathleen M. and Kathleen M. Kurz (1992) 'Women's nutrition through the life cycle: social and biological vulnerabilities', in Marge Koblinsky, Judith Timyan and Jill Gay (eds) *The Health of Women: A Global Perspective*, Boulder, CO: Westview Press, pp. 63–90.

Michael, Carmen, Rosalie Gardiner, Martin Prowse and Minu Hemmati (1999) 'Women's employment in tourism world-wide: data and statistics', in UNED-UK (eds) *Gender and Tourism: Women's Employment and Participation in Tourism*, Report for the United Nations Commission on Sustainable Development, 7th Session, April, London: DFID.

Mies, M. and V. Shiva (1993) *Ecofeminism*, London: Zed Books.

Miller, Judith (2003) 'US expands Afghan aid for maternal and child health', *New York Times*, 27 January: A13.

Miller, Vernice, Moya Hallstein and Susan Quass (1996) 'Feminist politics and environmental justice: women's community activism in West Harlem, New York', in Dianne Rocheleau, Barbara Thomas-Slayter and Ester Wangari (eds) *Feminist Political Ecology: Global Issues and Local Experiences*, London and New York: Routledge.

Mills, Beth (2002) 'Migration and land tenure in Carriacou, West Indies', unpublished Ph.D. dissertation, University of California, Davis.

Milwertz, Cecilia Nathansen (1997) *Accepting Population Control: Urban Chinese Women and the One-Child Family Policy*, Nordic Institute of Asian Studies, Monograph No. 74, Richmond, Surrey: Curzon Press.

Mohai, P. (1992) 'Men, women and the environment: an examination of the gender gap in environmental concern and activism', *Society and Natural Resources* 5 (1): 1–19.

Mohanty, Chandra Talpaty (1984) 'Under Western eyes: feminist scholarship and colonial discourses', *Boundary 2* 12 (3) and 13 (1): 333–58.

Molyneux, Maxine (1985) 'Mobilisation without emancipation: women's interests, state, and revolution in Nicaragua', *Feminist Studies* 11 (2): 227–54.

Momsen, Janet H. (1986) 'Migration and rural development in the Caribbean', *Tijdschrift voor Economische en Sociale Geografie* 77 (1): 50–8.

Momsen, Janet H. (1988a) 'Gender roles in Caribbean agricultural labour', in Malcolm Cross and Gad Heuman (eds) *Labour in the Caribbean*, Basingstoke: Macmillan.

Momsen, Janet H. (1988b) 'Changing gender roles in Caribbean peasant agriculture', in John S. Brierley and Hymie Rubenstein (eds) *Small Farming and Peasant Agriculture in the Caribbean*, Manitoba Geographical Studies 10, Winnipeg: University of Manitoba, pp. 83–100.

Momsen, Janet H. (1992a) 'Gender selectivity in Caribbean migration', in Sylvia Chant (ed.) *Gender and Migration in Developing Countries*, London: Belhaven Press, pp. 73–90.

Momsen, Janet H. (1992b) 'Regional patterns in the working lives of third world women', *Boletin de Geografica Teoretica* 22 (43–4): 292–5.

Momsen, Janet H. (1993) 'Gender and environmental perception in the eastern Caribbean', in D. G. Lockhart, David Drakakis-Smith and John Schrembi (eds) *The Development Process in Small Island States*, London: Routledge, pp. 57–70.

Momsen, Janet H. (ed.) (1999) *Gender, Migration and Domestic Service*, London and New York: Routledge.

Momsen, Janet H. (2000) 'Gender differences in environmental concern and perception', *Journal of Geography* 99: 47–56.

Momsen, Janet H. (2001) 'Backlash: or how to snatch failure from the jaws of success in gender and development', *Progress in Development Studies* 1 (1): 51–6.

Momsen, Janet H. (2002a) 'Myth or math: the waxing and waning of the female-headed household', *Progress in Development Studies* 2 (2): 145–51.

Momsen, Janet H. (2002b) 'Gender and entrepreneurship in post-communist Hungary', in Al Rainnie, Adrian Smith and Adam Swain (eds) *Work, Employment and Transition in Post-Communist Europe*, London: Routledge.

Momsen, Janet H. (2002c) 'NGOs, gender and indigenous grassroots development', *Journal of Development Studies* 14: 1–9.

Momsen, Janet H. and Vivian Kinnaird (eds) (1993) *Different Places, Different Voices: Gender and Development in Africa, Asia and Latin America*, London and New York: Routledge.

Moore, Molly (2001) 'Women victims of a village tradition', *Guardian Weekly*, 16–22 August: 29.

Moser, Caroline O. N. (1992) 'Adjustment from below: low-income women, time and the triple role in Guayaquil, Ecuador', in Haleh Afshar and Carolyne Dennis (eds) *Women and Adjustment Policies in the Third World*, Women's Studies at York, Macmillan Series, Basingstoke: Macmillan.

Moser, Caroline O. N. (1993) *Gender, Planning and Development: Theory, Practice and Training*, London and New York: Routledge.

Moser, Caroline O. N. and Fiona C. Clark (2001) 'Gender, conflict, and building sustainable peace: recent lessons from Latin America', *Gender and Development* 9 (3): 29–39.

Moser, Caroline O. N. and Fiona C. Clark (eds) (2001) *Victims, Perpetrators or Actors? Gender, Armed Conflict and Political Violence*, London and New York: Zed Books.

Moser, Caroline O. N. and Linda Peake (eds) (1987) *Women and Human Settlements and Housing*, London and New York: Tavistock Publications.

Mukherjee, Sanjukta and Todd Benson (2003) 'The determinants of poverty in Malawi, 1998', *World Development* 31 (2): 339–58.

Narayan, Deepa and Patti Petesch (eds) (2002) *Voices of the Poor from Many Lands*, New York: Oxford University Press for the World Bank.

Narayan, Deepa, Robert Chambers, Meera K. Shah and Patti Petesch (2000a) *Voices of the Poor: Crying out for Change*, New York: Oxford University Press for the World Bank.

Narayan, Deepa with Raj Patel, Kai Schafft, Anne Rademacher and Sarah Koch-Schulte (2000b) *Voices of the Poor: Can Anyone Hear Us?*, New York: Oxford University Press for the World Bank.

Nasreen, Mahbuba (2000) 'Coping mechanisms of rural women in Bangladesh during floods: a gender perspective', in Nasreen Ahmad and Hafiza Khatun (eds) *Disaster Issues and Gender Perspectives: Conference Proceedings*, Dhaka: Bangladesh Geographical Society, University of Dhaka, Bangladesh, pp. 311–24.

Navarro, Mireya (2002) 'Who is killing the young women of Cuidad Juárez? A filmmaker seeks answers', *New York Times*, 19 August: B3.

Nelson, Nici (1992) 'The women who have left and those who have stayed behind: rural–urban migration in central and western Kenya', in Sylvia Chant (ed.) *Gender and Migration in Developing Countries*, London: Belhaven Press.

Nesmith, Cathy and Sarah Radcliffe (1993) '(Re)mapping Mother Earth: a geographical perspective on environmental feminisms', *Environment and Planning D* 11: 379–94.

Nettles, Kimberley (1998) 'Beyond rum and corned beef politics: the development of 'interactional oppositional consciousness' in the mobilization of grassroots women in Guyana, South America', unpublished dissertation, University of California, Los Angeles; Ann Arbor: University Microfilms International.

New York Times (2002) 'Reuters report "Sierra Leone: 10 arrested in genital cutting"', 1 August: A6.

Noel, Claudel (2001) 'Scavenging and solid waste management in Port au Prince, Haiti', unpublished M.Sc. thesis, Department of Geography, University of the West Indies, Jamaica.

Norr, James L. and Kathleen F. Norr (1997) 'Women's status in peasant-level fishing', in Carolyn E. Sachs (ed) *Women Working in the Environment*, Washington, DC: Taylor & Francis.

Norris, Pippa (2001) 'Breaking the barriers: positive discrimination policies for women', in Jyette Klausen and Charles Maier (eds) *Has Liberalism Failed Women: Parity, Quotas and Political Representation*, New York: St Martin's Press.

Oakley, Emily (2002) Field notes.

Odame, Helen Hambly, Nancy Hafkin, Gesa Wesseler and Isolina Boto (2002) *Gender and Agriculture in the Information Society*, Briefing Paper 55, The

Hague: International Service for National Agricultural Research. Available on-line: http://www.isnar.cgiar.org (accessed 8 November 2002).

Olcott, Martha (2000) *Regional Study on Human Development and Human Rights: Central Asia*, Human Development Report 2000 Background Paper No. 52. Available on-line: http://www.hdr-undp.org/docs/publications (accessed 6 December 2002).

Oloka-Onyango, Joseph (2000) *Human Rights and Sustainable Development in Contemporary Africa: A New Dawn, or Retreating Horizons?*, Human Development Report 2000 Background Paper No.53. Available on-line: http://www.hdr.undp.org/docs/publications (accessed 6 December 2002).

Oxaal, Z. and S. Cook (1998) *Health and Poverty Gender Analysis*, BRIDGE Report No. 46, Brighton, Sussex: Institute of Development Studies.

Parpart, Jane (2002) 'Gender and empowerment: new thoughts, new approaches', in Vandana Desai and Robert B. Potter (eds) *The Companion to Development Studies*, London: Arnold.

Pearson, Maggie (1987) 'Old wives or young midwives? Women as caretakers of health: the case of Nepal', in Janet H. Momsen and Janet Townsend (eds) *Geography of Gender in the Third World*, London: Hutchinson; Albany: State University of New York, pp. 116–30.

Pearson, Ruth (1993) 'Gender and new technology in the Caribbean: new work for women?', in Janet H. Momsen (ed.) *Women and Change in the Caribbean*, London: James Currey; Indianapolis: Indiana University Press; Kingston, Jamaica: Ian Randle.

Pearson, Ruth (1998) 'Nimble fingers revisited: reflections on women and third world industrialization in the late twentieth century', in C. Jackson and R. Pearson (eds) *Feminist Visions of Development: Gender Analysis and Policy*, London and New York: Routledge.

Pearson, Ruth (2000) ' "Moving the goalposts": gender and globalization in the twenty-first century', in Caroline Sweetman (ed.) *Gender in the 21st Century*, Oxford: Oxfam GB.

Pearson, Ruth (2002) 'Gender evaluation of the Cut the Cost Campaign', *Links* July: 3.

Percival, Debra (2002) 'The World Summit on Sustainable Development', *Courier ACP-EU* 194: 6–12.

Phua, Lily and Brenda S. A. Yeoh (2002) 'Nine months: women's agency and the pregnant body in Singapore', in Brenda S. A. Yeoh, Peggy Teo and Shirlena Huang (eds) *Gender Politics in the Asia-Pacific Region*, London and New York: Routledge.

Pickford, J. A. (ed.) (2000) 'Looking back: participatory impact assessment', in J. A. Pickford (ed.) *Proceedings of the 26th WEDC Conference*, Loughborough: Water, Engineering and Development Centre, Loughborough University. Available on-line: http://www.id21.org (accessed 3 February 2003).

Pinnawala, Mallika (1996) 'The impact of changes in rice farming technology on seasonal migration of women', in A. Wickramasinghe (ed.) *Development Issues across Regions: Women, Land and Forestry*, Colombo: CORRENSA.

Plumwood, Val (1993) *Feminism and the Mastery of Nature*, London: Routledge.

Poonyarat, Chayanit (2002) 'Male sex workers face HIV risks, but get less attention', InterPress Service News Agency. Available on-line: http://www.194.183.22.100/ips (accessed 25 November 2002).

Porter, Marilyn and Ellen Judd (eds) (1999) *Feminists Doing Development*, London and New York: Zed Books.

Potter, R. and Lloyd-Evans, Sally (1998) *The City in the Developing World*, Harlow: Longman.

PRB (Population Reference Bureau) (2002) *Women of Our World*, New York: PRB.

Presser, Harriet B. (2000) 'Comment: a gender perspective for understanding low fertility in post-transition societies', in Rudolpho Bulatao and John B. Casterline (eds) *A Global Fertility Transition*, New York: Population Council. (A supplement to *Population and Development Review* 27: 177–83.)

Pryor, J. (1987) 'Production and reproduction of malnutrition in an urban slum in Khulna, Bangladesh', in Janet H. Momsen and Janet Townsend (eds) *Geography of Gender in the Third World*, London: Hutchinson.

Quarrie, Joyce (1992) *Earth Summit 1992*, London: The Regency Press Corporation.

Radcliffe, Sarah A. and Sallie Westwood (1993) *'Viva': Women and Popular Protest in Latin America*, London and New York: Routledge.

Rai, Saritha (2002) 'India is regaining contracts with the US: ranks no. 1 in outsourcing deals', *New York Times*, 25 December: W1 and W7.

Raju, Saraswati, Peter J. Atkins, Naresh Kumar and Janet Townsend (1999) *Atlas of Women and Men in India*, New Delhi: Kali for Women.

Rama, Martín (2002) 'The gender implications of public sector downsizing: the reform program of Vietnam', *World Bank Research Observer* 17 (2): 167–89.

Ramamurthy, Priti (1997) 'Rural women and irrigation: patriarchy, class, and the modernizing state in South India', in Carolyn E. Sachs (ed.) *Women Working in the Environment*, Washington, DC: Taylor & Francis.

Randriamaro, Zo (2001) 'Financing for women's economic rights: how to escape the micro-finance ghetto in Africa?' *AWID News* 15 (1): 8–9, 14.

Ransome, Pamela (2001) 'Women, pesticides and sustainable agriculture'. Available on-line: http://www.earthsummit2002.org/wcaucus (accessed 10 November 2002).

Rashid, Haroun er (2000) 'Disaster: issues and gender perspectives', in Nasreen Ahmad and Hafiza Khatun, *Disaster Issues and Gender Perspectives*: *Conference Proceedings*, Dhaka: Bangladesh Geographical Society, University of Dhaka, Bangladesh: pp. xxiii–xxxvii.

Rathgeber, E. M. (1990) 'WID, WAD, GAD: trends in research and practice', *The Journal of Developing Areas* 24 (1): 489–502.

Rehn, Elisabeth and Ellen Johnson Sirleaf (2002) *Women, War and Peace: The Independent Expert's Assessment on the Impact of Armed Conflict on Women and Women's Role in Peace-building*, New York: Unifem.

Rivers, J. (1987) 'Women and children last: an essay on sex discrimination', *Disasters* 6 (4): 256–67.

Roberts, Dan and James Kynge (2003) 'How cheap labour, foreign investment and rapid industrialisation are creating a new workshop of the world', *Financial Times*, 4 February: 13.

Rocheleau, Dianne, Barbara Thomas-Slayter and Ester Wangari (eds) (1996) *Feminist Political Ecology: Global Issues and Local Experiences*, London and New York: Routledge.

Roman, Denise (2001) 'Gendering Eastern Europe: pre-feminism, prejudice and East–West dialogues in post-communist Romania', *Women's Studies International Forum* 24 (1): 53–66.

Rowlands, J. (1997) *Questioning Empowerment: Working with Women in Honduras*, Oxford: Oxfam Publications.

Rumi, Syed Rafiqul Alam and Sk. Ohiduzzaman (2000) 'Impact of biomass fuel crisis on rural women and children: a case study of four villages', in Nasreen Ahmad and Hafiza Khatun (eds) *Disaster Issues and Gender Perspectives*: *Conference Proceedings*, Dhaka: Bangladesh Geographical Society, University of Dhaka, Bangladesh, pp. 194–203.

Sachs, Carolyn E. (ed.) (1997) *Women Working in the Environment*, London and Washington, DC: Taylor & Francis.

Sachs, Carolyn E., Kishor Gajurel and Mariela Bianco (1997) 'Gender, seeds and biodiversity', in Carolyn E. Sachs (ed.) *Women Working in the Environment*, Washington, DC: Taylor & Francis.

Saferworld (2002) Available on-line: http://www.saferworld.co.uk (accessed 1 November 2002).

Sahn, David E. and David C. Stifel (2003) 'Progress toward the Millennium Development Goals in Africa', *World Development* 31 (1): 23–52.

Salomon, Joshua A. and Christopher J. L. Murray (2002) 'The epidemiologic transition revisited: compositional models for causes of death by age and sex', *Population and Development Review* 28 (2): 205–28.

Samarasinghe, Vidyamali (1993) 'Access of female plantation workers of Sri Lanka to basic needs provision', in Janet H. Momsen and V. Kinnaird (eds) *Different Places, Different Voices: Gender and Development in Africa, Asia and Latin America*, London and New York: Routledge.

Sanez, Nair Carrasco, Rosa Maria Door de Ubillas and Irma Salvatierra Guillen with Sebastiao Mendonca Ferreira (1998) *Increasing Women's Involvement in Community Decision Making: A Means to Improve Iron Status (Peru)*, Research Report Series No. 1, Washington, DC: International Center for Research on Women (ICRW).

Sarin, M. (1995) 'Regenerating India's forests: reconciling gender equity with joint forest management', *IDS Bulletin* 26 (1): 83–91.

Sass, J. and L. Ashford (2002) *2002 Women of Our World*, Washington, DC: Population Reference Bureau. Available on-line: http://www.prb.org (accessed 2 August 2002).

Satheesh, P. V. (2000) 'Linking to community development: using participatory approaches to *in situ* conservation', in Esbern Friss-Hansen and Bhuwon R.

Sthapit (eds) *Participatory Approaches to the Conservation and Use of Plant Genetic Resources*, Rome: IPGRI.

Scott, A. McEwan (1986) 'Economic development and urban women's work: the case of Lima, Peru' in R. Anker and C. Hein (eds) *Sex Inequalities in Urban Employment in the Third World*, London: Macmillan.

Seager, Joni (1996) 'Rethinking the environment: women and pollution', *Political Environments* 3: 14–16.

Seager, Joni (1997) *The State of Women in the World Atlas*, 2nd edition, London: Penguin Books.

Seelock, Nadiya (2001) *Salvaging in Port of Spain, Trinidad*, unpublished M.Sc. thesis, Department of Geography and Geology, University of the West Indies, Jamaica.

Sen, A. (1990) 'More than a 100 million women are missing', *New York Review of Books*, 20 December.

Sen, G. and C. Grown (1987) *Development Crises, and Alternative Visions: Third World Women's Perspectives*, New York: Monthly Review Press.

Sengupta, Somini (2002) 'Child traffickers prey on Bangladesh', *New York Times*, 29 April: A6.

Shah, M. K. and P. Shah (1995) 'Gender, environment and livelihood security: an alternative viewpoint from India', *IDS Bulletin* 26 (1): 75–82.

Shamin, Ishrat (1992) 'Slavery and traffic in women and children: recent trends in Bangladesh', in Latifa Akanda (ed.) *Eshon: Collected Articles 1992*, Dhaka: Women for Women, a Research and Study Group, pp. 72–83.

Shiva, Vandana (1989) *Staying Alive: Women, Ecology and Development*, London: Zed Books.

Silovic, Darko (1999) *Regional Study on Human Development and Human Rights in Central and Eastern Europe*, Human Development Report Background Paper 1999. Available on-line: http://www.hdr.undp.org/docs/publications (accessed December 2002).

Silvey, Rachel (1998) ' "Ecofeminism" in geography', *Ethics, Place and Environment* 1 (2): 243–51.

Simard, Paule and Maria De Koninck (2001) 'Environment, living spaces, and health: compound-organization practices in a Bamako squatter settlement, Mali', *Gender and Development* 9 (2): 28–39.

Simons, Marlise (2003) 'Dutch appear to re-elect conservative leader in mixed outcome', *New York Times*, 23 January: A7.

Smitasiri, Suttiak and Sakorn Dhanamitta (1999) *Sustaining Behavior Change to Enhance Micronutrient Status: Community- and Women-based Interventions in Thailand*, Research Report No.2, Washington, DC: International Center for Research on Women (ICRW).

Smith, Garrett C. (1995) 'Food and diet in southern Burkina Faso: nutritional overview employing a Rapid Assessment Procedure (RAP) and the carotenoid and mineral analyses of selected wild and cultivated edible plants', unpublished Ph.D. dissertation, University of California, Davis.

Smith, Lisa and Lawrence Haddad (1999) *Explaining Child Malnutrition in Developing Countries: A Cross-Country Analysis*, Discussion Paper Briefs, Discussion Paper 60, Washington, DC: IFPRI. Available on-line: http//www.cgiar.org/ifpri (accessed 9 July 1999).

Smith, Mogha Kamal (2002) 'Gender, poverty, and intergenerational vulnerability to HIV/AIDS', *Gender and Development* 10 (3): 63–70.

Songsore, Jacob (1999) 'Urban environment and human well-being: the case of the Greater Accra Metropolitan Area, Ghana', *Bulletin of the Ghana Geographical Association* 21 (July): 1–10.

Sri Lanka Department of Census and Statistics (1987) *Census, 1981*, Colombo: Department of Census and Statistics.

Statistics Iceland (2002) *Iceland in Figures, Vol. 7*, ed. Bjorgvin Sigurosson, Reykjavik: Hagsofa Press. Available on-line: http://www.statice.is (accessed 19 August 2002).

Sturgeon, Noel (1997) *Ecofeminist Natures: Race, Gender, Feminist Theory and Political Action*, New York: Routledge.

Sunday Observer (2002) 'Brides', 29 September. Available on-line: http://www.sundayobserver.lk (accessed 4 October 2002).

Swain, M. B. and Janet H. Momsen (eds) (2002) *Gender/Tourism/Fun(?)*, New York: Cognizant Communication Corporation.

Sweetman, Caroline (ed.) (1998) *Gender and Migration*, Oxford: Oxfam GB.

Sweetman, Caroline (2000) 'Editorial', in Caroline Sweetman (ed.) *Gender in the 21st Century*, Oxford: Oxfam GB.

Szorenyi, Iren Kukorelli (2000) Personal communication.

Tam, Vicky C. W. (1999) 'Foreign domestic helpers in Hong Kong and their role in childcare provision' in Janet H. Momsen (ed.) *Gender, Migration and Domestic Service*, London and New York: Routledge.

Tapia, Mario E. and Ana de la Torre (1997) *Women Farmers and Andean seeds*, Rome: IPGRI.

Teo, Peggy and Brenda S. A. Yeoh (1999) 'Interweaving the public and the private: women's responses to population policy shifts in Singapore', *International Journal of Population Geography* 5: 79–96.

Thomas-Hope, Elizabeth (ed.) (1998) *Solid Waste Management: Critical Issues for Developing Countries*, Kingston, Jamaica: Canoe Press, University of the West Indies.

Tinker, Anne, Patricia Daly, Cynthia Green, Helen Saxenian, Rama Lakshminarayanan and Kirron Gill (1994) *Women's Health and Nutrition: Making a Difference*, World Bank Discussion Paper 256, Washington, DC: World Bank.

Townsend, Janet and Sally Wilson d'Acosta (1987) 'Gender roles in colonization of rainforest: a Colombian case study', in Janet H. Momsen and Janet Townsend (eds) *Geography of Gender in the Third World*, London: Hutchinson.

Tsegaye, B. (1997) 'The significance of biodiversity for sustaining agricultural production and role of women in the traditional sector: the Ethiopian experience', *Agriculture, Ecosystems and Environment* 62: 215.

Tsui, Amy Ong (2000) 'Population policies, family planning and fertility: the record', in Rudolpho A. Bulatao and John B. Casterline (eds) *Global Fertility Transition*, New York: Population Council. (A supplement to *Population and Development Review* 27: 184–204.)

Ulluwishewa, R. (1993) *Development Planning, Environmental Degradation and Women's Fuelwood Crisis: A Sri Lankan Case Study*, Working Paper 28, Davis, CA: IGU Commission on Gender and Geography.

Ulluwishewa, R. (1997) 'Rural development and its consequences for women: a case study of a development project in Sri Lanka', *Geography* 82 (355): 110–17.

UNAIDS (2000) *Men and AIDS: A Gendered Approach*, Geneva: UNAIDS. Available on-line: http://www.unaids.org (accessed 11 February 2003).

UNAIDS (2002) *Report on the Global HIV/AIDS Epidemic 2002*, Geneva: United Nations. Available on-line: http://www.unaids.org (accessed 15 July 2002).

UNDP (United Nations Development Programme) (1999) *Human Development Report 1999*, New York and Oxford: Oxford University Press.

UNDP (United Nations Development Programme) (2001) *Human Development Report 2001*, New York: United Nations.

UNDP (United Nations Development Programme) (2002) *Human Development Report 2002*. Available on-line: http://hdr.undp.org (accessed 22 November 2002).

UNDP (United Nations Development Programme) (2003) *Human Development Report 2003. Millennium Development Goals: A Compact among Nations to End Human Poverty*. Available on-line: http://www.undp.org/hdr2003 (accessed 16 July).

UNICEF (1999) *Women in Transition: The MONEE Project*, CEE/CIS/Baltics, Regional Monitoring Report No. 6, Florence, Italy: International Child Development Centre.

United Nations (1995a) 'Program of action of the 1994 International Conference on Population and Development (Chapters I–VIII), *Population and Development Review* 21 (1): 187–213.

United Nations (1995b) *The World's Women 1995: Trends and Statistics*, New York: United Nations.

United Nations (1996) *The United Nations and the Advancement of Women, 1945–1996*, New York: United Nations, Department of Public Information.

United Nations (2000) *The World's Women 2000: Trends and Statistics*, New York: United Nations.

United Nations (2001) *Human Development Indicators*, New York: United Nations.

United Nations (2002) 'A fact sheet on women and armed conflict', *Women's Global Net #212*, 23 October. Available on-line: http://www.iwtc.org (accessed 1 November 2002).

USAID (1999) Women as chattel: the emerging global market in trafficking, *Gender Matters Quarterly* 1 (February).

USAID (2001) 'Gender and community conservation: part 1', *Gender Matters Quarterly* 3 (June). Available on-line: http://www.usaid.gov/wid/pubs/ (accessed 19 November 2001).

Varley, Ann (2002) 'Gender, families and households', in Vandana Desai and Robert B. Potter (eds) *The Companion to Development Studies*, London: Arnold.

Vidal, John (1989) 'Root causes, root answers', *Guardian*, 3 November.

Walker, Barbara (1998) 'Sisterhood and seine-nets: engendering the development, science and conservation of Ghana's marine fishery', unpublished Ph.D. dissertation, University of California, Berkeley.

Walker, Anne S. and Karen Banks (2003) 'World summit on the information society (WSIS): update on plans and preparations including gender input', *Women's GlobalNet #215*, 13 January. Available on-line: http://www.iwtc. org (accessed 25 January 2003).

Warren, Karen J. (1990) 'The power and the promise of ecological feminism, *Environmental Ethics* 12 (2): 125–46.

Wastl-Walter, Doris (1996) 'Protecting the environment against state policy in Austria: from women's participation in protest to new voices in Parliament', in Dianne Rocheleau, Barbara Thomas-Slayter and Ester Wangari (eds) *Feminist Political Ecology: Global Issues and Local Experiences*, London and New York: Routledge.

Waylen, Georgina (1996) *Gender in Third World Politics*, Buckingham: Open University Press.

Weiss, Ellen and Geeta Rao Gupta (1998) Bridging the Gap: Addressing Gender and Sexuality in HIV Prevention, Washington, DC: ICRW.

Wennerholm, Caroline Johansson (2002) 'Crossing borders and building bridges: the Baltic Region Networking Project', *Gender and Development* 10 (1): 10–19.

Wickramasinghe, Anoja (1993) 'Women's roles in rural Sri Lanka', in Janet H. Momsen and Vivian Kinnaird (eds) *Different Places, Different Voices: Gender and Development in Africa, Asia and Latin America*, London and New York: Routledge.

Wickramasinghe, Anoja (1994) *Deforestation, Women and Forestry*, Amsterdam: Institute for Development Research.

Wilkinson, C. R. (1987) 'Women, migration and work in Lesotho', in Janet H. Momsen and Janet Townsend (eds) *Geography of Gender in the Third World*, Albany, NY: State University of New York Press; London: Hutchinson, pp. 117–31.

Wilkinson, C. R. (1989) 'Migration in Lesotho: a study of population movements in a labour reserve economy', paper presented at the Commonwealth Geographical Bureau Workshop on Gender and Development, Newcastle upon Tyne, England, 16–21 April 1989.

Williams, Suzanne and Rachel Masika (2002) 'Editorial', *Gender and Development* 10 (1): 2–9.

Winkler, Edwin A. (2002) 'Chinese reproductive policy at the turn of the millennium', *Population and Development Review* 28 (3): 379–418.

Wisner, Ben (1993) 'Disaster vulnerability: scale, power and daily life', *GeoJournal* 30 (2): 127–40.

Wolf, Diane L. (1992) *Factory Daughters: Gender, Household Dynamics and Rural Industrialization in Java*, Berkeley: University of California Press.

Women in Action (2001) 'Transforming Filipino men through innovative projects', *Women in Action* 1. Available on-line: http://www.isiswomen.org/pub/wia/wia1 (accessed 3 November 2001).

World Bank (1989) *Development Report 1989*, Washington, DC: World Bank.

World Bank (1993) *World Development Report 1993: Investing in Health*, New York: Oxford University Press.

World Bank (2001) *World Development Indicators*, Washington, DC: World Bank.

World Organization Against Torture (Organization Mondiale Contre la Torture, OMCT) (2001) *Violence Against Women: For the Protection and Promotion of the Human Rights of Women*, Ten Reports/Year 2000, Geneva: OMCT. Summary available e-mail: development-gender@yahoogroups.com (accessed 18 May 2001).

Yeoh, B. S. A. and S. Huang (1999) 'Singapore women and foreign domestic workers: negotiating domestic work and motherhood', in Janet H. Momsen (ed.) *Gender, Migration and Domestic Service*, London and New York: Routledge.

Yinger, Nancy V. (1998) *Unmet Need for Family Planning: Reflecting Women's Perceptions*, Washington, DC: International Center for Research on Women.

Young, Kate (2002) 'WID, GAD and WAD', in Vandana Desai and Robert B. Potter (eds) *The Companion to Development Studies*, London: Arnold.

Zimmerer, Karl S. (1991) 'Seeds of peasant subsistence: agrarian structure, crio ecology and Quechua agriculture in reference to the loss of biological diversity in northern Peruvian Andes', Ph.D. dissertation, University of California, Berkeley; Ann Arbor, Michigan: University Microfilms International.

Index